Elements of Algebraic Coding

ELEMENTS OF ALGEBRAIC CODING SYSTEMS

VALDEMAR CARDOSO DA ROCHA JR., PH.D.
Federal University of Pernambuco
Brazil

MOMENTUM PRESS

MOMENTUM PRESS, LLC, NEW YORK

Elements of Algebraic Coding Systems
Copyright © Momentum Press®, LLC, 2014.

First published by Momentum Press®, LLC
222 East 46th Street, New York, NY 10017
www.momentumpress.net

ISBN-13: 978-1-60650-574-8 (print)
ISBN-13: 978-1-60650-575-5 (e-book)

Momentum Press Communications and Signal Processing Collection

DOI: 10.5643/9781606505755

Cover design by Jonathan Pennell
Interior design by Exeter Premedia Services Private Ltd., Chennai, India

10 9 8 7 6 5 4 3 2 1

Printed in the United States of America

To my daughter
Cynthia and to my son
Leandro

Contents

Abstract

Leaving behind 14 years of chaotic life in Brazil, I went to Switzerland with my family in August 1990. We spent a year and a half in Zurich, where I worked at the Federal Institute of Technology Zurich (ETHZ) with Prof. James L. Massey and interacted with his doctoral students and other members of the unit called Institut für Signal- und Informationsverarbeitung (ISI). Back in Brazil, this interaction continued and led to some joint work.

Since my return to Brazil I have been teaching error-correcting codes, information theory, and cryptography at the Federal University of Pernambuco.

This book serves as an introductory text to algebraic coding theory. The contents are suitable for final year undergraduate and first year graduate courses in electrical and computer engineering, and will give the reader knowledge of coding fundamentals that is essential for a deeper understanding of state-of-the-art coding systems. This book will also serve as a quick reference for those who need it for specific applications, like in cryptography and communications. Eleven chapters cover linear error-correcting block codes from elementary principles, going through cyclic codes and then covering some finite field algebra, Goppa codes, algebraic decoding algorithms, and applications in public-key cryptography and secret-key cryptography. At the end of each chapter a section containing problems and solutions is included. Three appendices cover the Gilbert bound and some related derivations, a derivation of the MacWilliams' identities based on the probability of undetected error, and two important tools for algebraic decoding, namely, the finite field Fourier transform and the Euclidean algorithm for polynomials.

Keywords

codes, BCH codes, Goppa codes, decoding, majority logic decoding, time domain decoding, frequency domain decoding, Finite fields, polynomial factorization, error-correcting codes, algebraic codes, cyclic.

Acknowledgments

This book is the result of work done over many years in the Department of Electronics and Systems of the Federal University of Pernambuco. Collaboration with Brazilian colleagues at the Federal University of Campina Grande, University of Campinas and Pontifical Catholic University of Rio de Janeiro is gratefully acknowledged. The author is grateful to all members of the Communications Research Group at the Federal University of Pernambuco, for their contributions in varied forms, including seminars and informal discussions, as well as to his colleagues at the Institute for Advanced Studies in Communications. The author is also grateful to his friends in the United Kingdom, Professor Paddy Farrell, Professor Mike Darnell, Professor Bahram Honary, and Professor Garik Markarian for a long lasting collaboration through Lancaster University and the University of Leeds. And last, but not least, the author wishes to take this opportunity to acknowledge the extremely useful experience of learning more about coding and cryptography through many little discussions with Jim Massey.

Finally, the author wishes to thank the Collection Editor Orlando Baiocchi; and Shoshanna Goldberg, Millicent Treloar, and Jeff Shelstad, from Momentum Press, for their strong support of this publication project from the very start and for helping with the reviewing process.

Chapter 1

BASIC CONCEPTS

1.1 Introduction

Reliable data transmission at higher data rates, has always been a constant challenge for both engineers and researchers in the field of telecommunications. Error-correcting codes (Lin and Costello Jr. 2004), have without doubt contributed in a significant way for the theoretical and technological advances in this area. Problems related to storage and recovery of large amounts of data in semiconductor memories (Chen and Hsiao 1984), have also been benefited from error-correcting coding techniques.

Another important aspect related to both data transmission and data storage concerns data security and data authenticity. However, security and authenticity belong to the study of cryptology (Konheim 1981) and will not be considered by us in this book. Transmission reliability, referred to earlier, concerns only the immunity of the transmitted data to noise and other types of interference, ignoring the possibility of message interception by a third party.

Frequently in the context of digital communications, we face problems of detection and correction of errors caused by noise during transmission, or that have affected stored data. Situations of this kind occur, for example, in a banking data transmission network where, ideally, errors should never occur.

Digital communication systems keep changing their appearance as far as circuits and components are concerned, as a consequence of changes in technology. For example, old communication systems evolved from the electromechanical relay to thermionic valves, later to transistors, and so on. Even so a closer look at such systems reveals that all of them can be represented by a block diagram as shown in Figure 1.1, the blocks of which are defined as follows.

- **Source:** The originator of information to be transmitted or stored. As examples of information sources we mention the output of a computer terminal, the output of a microphone or the output of a remote sensor in a telemetry system. The source is often modeled as a stochastic process or as a random data generator.

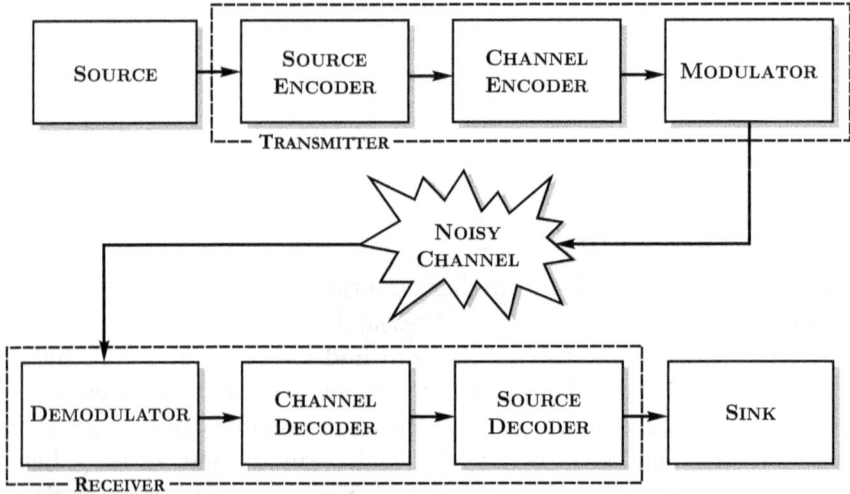

Figure 1.1. Digital communication system.

- **Transmitter:** A transmitter converts the source output into waveforms appropriate for transmission or storage. The role of a transmitter can be subdivided as follows:

  (1) **Source encoder:** Very often a source encoder consists of just an analog to digital converter. For more sophisticated applications, a source encoder may perform the task of removing unnecessary detail from the data as, for example, in image processing.

  (2) **Channel encoder:** A channel encoder adds controlled redundancy to the data at the output of the source encoder to combat channel noise.

  (3) **Modulator:** A modulator translates the channel encoder output to a waveform appropriate for transmission or storage.

- **Channel:** A channel in practice is the physical media through which information passes before reaching the receiver. A channel may consist for example of a pair of wires, a microwave radio link, etc. Data traveling through the channel is subjected to noise in the form of undesirable disturbances which are, in certain ways, unpredictable. As a result of corruption by channel noise part of the information may be lost or severely mutilated. To predict or to measure the performance in a communication system it is necessary to characterize the channel noise mathematically by means of tools from statistics. In other words, it is necessary to mathematically model the channel.

■ **Receiver:** The role of a receiver in a communication system is to process the noisy channel output, aiming to detect the transmitted waveform and recover the transmitted data. The receiver part is normally the most complex part of a communication system and can be subdivided as follows:

(1) **Demodulator:** The demodulator processes the waveform received from the channel and delivers either a discrete (i.e., a quantized) output or a continuous (i.e., an unquantized) output to the channel decoder.

(2) **Channel decoder:** By operating on the demodulator output and applying decoding techniques the channel decoder attempts correction of possible errors and erasures before delivering its estimate of the corresponding source encoder output digits. The correction of errors is usually more complex than the correction of erasures since the positions of the latter are known to the decoder.

(3) **Source decoder:** A source decoder processes the channel decoder output, replacing redundancy removed earlier at the source encoder, and thus reconstituting the message to be delivered to the data sink.

■ **Sink:** A sink is the final recipient of the information transmitted. A sink can be, for example, a human being at the end of a telephone line, or a computer terminal.

1.2   Types of errors

Due to the presence of noise, as mentioned earlier, errors may occur during transmission or storage of data. These errors can occur sporadically and independently, in which case they are referred to as *random errors*, or else errors can appear in bursts of many errors each time it occurs, and are called *burst errors*, in which case the channel is said to have a memory.

1.3   Channel models

As mentioned earlier data traveling through the channel is corrupted by noise. Ideally the receiver should be able to process a continuous signal received from the channel. This situation is modeled by a channel with a discrete input and a continuous output. For practical reasons very often the receiver output needs to be quantized into a finite number of levels, typically 8 or 16 levels, which situation is modeled by a discrete channel. Two typical discrete channel models are the binary symmetric channel (BSC) and the binary erasure channel (BEC) (Peterson and Weldon Jr. 1972, pp.7–10). The BSC and the BEC somehow model two extreme situations, since each binary digit at the BSC output is either correct or assumes its complementary value (i.e., is in error), while the BEC outputs are either correct binary digits or are erased digits.

1.4   Linear codes and non-linear codes

Linear error-correcting codes are those codes for which the parity-check digits in a codeword result from a linear combination involving information digits. For nonlinear codes, on the other hand, parity-check digits may result from nonlinear logical operations on the information digits of a codeword, or else result from nonlinear mappings over a given finite field or finite ring, of linear codes over a finite field or finite ring of higher dimension (Hammons Jr. et al. 1994).

In the sequel only linear codes will be addressed due to their practical importance and also for concentrating, essentially in all publications, on error-correcting codes.

1.5   Block codes and convolutional codes

Depending on how the digits of redundancy are appended to the digits of information, two different types of codes result. Codes for which redundancy in a block of digits checks the occurrence of errors only in that particular block are called *block codes*. Codes where the redundancy in a block checks the presence or absence of errors in more than one block are called *convolutional codes*. Convolutional codes are a special case of tree codes (Peterson and Weldon Jr. 1972, p.5) which are important in practice but are not the subject of this book. Block and convolutional codes are competitive in many practical situations. The final choice of one depends on the factors such as data format, delay in decoding, and system complexity necessary to achieve a given error rate, etc.

No matter how well designed it is, any communication system is always disturbed by the action of noise, i.e., messages in its output can contain errors. In some situations it is possible that a long time will pass without the occurrence of errors, but eventually some errors are likely to happen. However, the practical problem in coding theory is not the provision of error-free communications but the design of systems that have an error rate sufficiently low for the user. For example, an error rate of 10^{-4} for the letters of a book is perfectly acceptable while that same error rate for the digits of a computer operating electronic funds transfer would be disastrous.

The maximum potential of error-correcting codes was established in 1948, with the Shannon coding theorem for a noisy channel. This theorem can be stated as follows.

Theorem. For any memoryless channel, whose input is a discrete alphabet, there are codes with information rate R (nats/symbol), with codewords of length n digits, for which the probability of decoding error employing maximum likelihood is bounded by $P_e < e^{-nE(R)}$, where $E(R) > 0$, $0 \leq R < C$, is a decreasing convex-$\cup$ function, specified by the channel transition probabilities and C is the channel capacity (Viterbi and Omura 1979, p.138).

The coding theorem proves the existence of codes that can make the probability of erroneous decoding very small, but gives no indication of how to construct such codes. However, we observe that P_e decreases exponentially when n is increased, which usually entails an increase in system complexity.

The goals of coding theory are basically:

- Finding long and efficient codes.

- Finding practical methods for encoding and efficient decoding.

Recent developments in digital hardware technology have made the use of sophisticated coding procedures possible, and the corresponding circuits can be rather complex. The current availability of complex processors makes the advantages derived from the use of codes become even more accessible.

1.6   Problems with solutions

(1) Suppose a source produces eight equally likely messages which are encoded into eight distinct codewords as 0000000, 1110100, 1101001, 1010011, 0100111, 1001110, 0011101, 0111010. The codewords are transmitted through a BSC with probability of error p, $p < 1/2$. Calculate the probability that an error pattern will not be detected at the receiver.

Solution: An error pattern will not be detected if the received word coincides with a codeword. For the set of codewords given, notice that the modulo 2 bit by bit addition of any two codewords produces a valid codeword. Therefore, we conclude that if an error pattern coincides with a codeword it will not be detected at the receiver. The probability of undetected error is thus: $(1 - p)^7 + 7p^4(1 - p)^3$.

(2) The set of codewords in the previous problem allows the correction of a single error in a BSC. Calculate the block error rate after decoding is performed on a received word.

Solution: The probability of error in a word after decoding, denoted as P_B, is equal to the probability of occurring i errors, $2 \leq i \leq 7$. Instead of computing P_B with the expression $\sum_{i=2}^{7} p^i(1 - p)^{7-i}$, it is simpler to compute $P_B = 1 - (1 - p)^7 - 7p(1 - p)^6$.

Chapter 2

BLOCK CODES

2.1 Introduction

Block codes can be easily characterized by their encoding process. The process of encoding for these codes consists in segmenting the message to be transmitted in blocks of k digits and appending to each block $n - k$ redundant digits. These $n - k$ redundant digits are determined from the k-digit message and are intended for just detecting errors, or for detection and correction of errors, or correcting erasures which may appear during transmission.

Block codes may be linear or nonlinear. In linear codes, as mentioned in Chapter 1, the redundant digits are calculated as linear combinations of information digits. Linear block codes represent undoubtedly the most well-developed part of the theory of error-correcting codes. One could say that this is in part due to the use of mathematical tools such as linear algebra and the theory of finite fields, or *Galois fields*. Due to their importance in practice, in what follows we mostly consider binary linear block codes, unless indicated otherwise. In general, the code alphabet is q-ary, where q denotes a power of a prime. Obviously for binary codes we have $q = 2$. A q-ary (n, k, d) *linear block code* is defined as follows.

DEFINITION 2.1 *A q-ary (n, k, d) linear block code is a set of q^k q-ary n-tuples, called codewords, where any two distinct codewords differ in at least d positions, and the set of q^k codewords forms a subspace of the vector space of all q^n q-ary n-tuples.*

The code rate R, or code efficiency, is defined as $R = k/n$.

2.2   Matrix representation

The codewords can be represented by vectors with n components. The components of these vectors are generally elements of a finite field with q elements, represented by $\mathrm{GF}(q)$, also called a *Galois field*. Very often we use the binary field, the elements of which are represented by 0 and 1, i.e., $\mathrm{GF}(2)$. As already mentioned, a linear code constitutes a subspace and thus any codeword can be represented by a linear combination of the basis vectors of the subspace, i.e., by a linear combination of linearly independent vectors. The basis vectors can be written as rows of a matrix, called the code *generator matrix* (Lin and Costello Jr. 2004, p.67).

Given a generator matrix $\mathbf{G}$ of a linear code with k rows and n columns, we can form another matrix $\mathbf{H}$, with $n-k$ rows and n columns, such that the row space of $\mathbf{G}$ is orthogonal to $\mathbf{H}$, that is, if $\mathbf{v}_i$ is a vector in the row space of $\mathbf{G}$ then

$$\mathbf{v}_i \mathbf{H}^T = \mathbf{0}, \ 0 \le i \le 2^k - 1.$$

The $\mathbf{H}$ matrix is called the code *parity-check matrix* and can be represented as

$$\mathbf{H} = [\mathbf{h} : I_{n-k}],$$

where $\mathbf{h}$ denotes an $(n-k) \times k$ matrix and I_{n-k} is the $(n-k) \times (n-k)$ identity matrix. It is shown, e.g., in (p.69), that the $\mathbf{G}$ matrix can be written as

$$\mathbf{G} = [I_k : \mathbf{g}], \tag{2.1}$$

where $\mathbf{g}$ denotes a $k \times (n-k)$ matrix and I_k denotes the $k \times k$ identity matrix. The form of $\mathbf{G}$ in (2.1) is called the reduced echelon form (Peterson and Weldon Jr. 1972, p.45–46). The $\mathbf{g}$ and $\mathbf{h}$ matrices are related by the expression $\mathbf{g} = \mathbf{h}^\mathrm{T}$. Since the rows of $\mathbf{H}$ are linearly independent, they generate a $(n, n-k, d')$ linear code called the *dual code* of the (n, k, d) code generated by $\mathbf{G}$. The code $(n, n-k, d')$ can be considered as the dual space of the (n, k, d) code generated by $\mathbf{G}$.

Making use of the matrix representation, we find that an encoder has the function of performing the product $\mathbf{mG}$ of a row matrix $\mathbf{m}$, with k elements which represent the information digits, by the $\mathbf{G}$ matrix. The result of such an operation is a linear combination of the rows of $\mathbf{G}$ and thus a codeword.

2.3   Minimum distance

The ability of simply detecting errors, or error detection and error correction of a code is directly linked to a quantity, defined later, that is its *minimum distance*. Before doing that however we define *Hamming weight* of a vector and the *Hamming distance* between two vectors.

DEFINITION 2.2 *The Hamming weight $W_H(\mathbf{v})$ of a vector $\mathbf{v}$ is the number of nonzero coordinates in $\mathbf{v}$.*

DEFINITION 2.3 *The Hamming distance $d_H(\mathbf{v}_1, \mathbf{v}_2)$ between two vectors, $\mathbf{v}_1$ and $\mathbf{v}_2$, having the same number of coordinates, is the number of positions is which these two vectors differ.*

We observe that the Hamming distance is a metric (p.17).

DEFINITION 2.4 *The minimum distance of a code is the smallest Hamming distance between pairs of distinct codewords.*

Denote by $q = p^m$ the cardinality of the code alphabet, where p is a prime number and m is a positive integer. Due to the linearity property, the modulo-q sum of any two codewords of a linear code results in a codeword. Suppose that $\mathbf{v}_i, \mathbf{v}_j$ and $\mathbf{v}_l$ are codewords such that $\mathbf{v}_i + \mathbf{v}_j = \mathbf{v}_l$. From the definitions of Hamming distance and Hamming weight it follows that $d_H(\mathbf{v}_i, \mathbf{v}_j) = W_H(\mathbf{v}_l)$. Hence we conclude that, to determine the minimum distance of a linear code means to find the minimum nonzero Hamming weight among the codewords. This last remark brings a great simplification to computing the minimum distance of a linear code because if the code has M codewords, instead of making C_M^2 operations of addition modulo-q and the corresponding Hamming weight calculation, it is sufficient to calculate the Hamming weight of the $M - 1$ nonzero codewords only. In special situations where the code, besides linearity, has an additional mathematical structure, the determination of the minimum distance, or the determination of upper or lower bounds for the minimum distance can be further simplified.

In a code with minimum distance d, the minimum number of changes necessary to convert a codeword into another codeword is at least d. Therefore, the occurrence of up to $d - 1$ errors per codeword during a transmission can be detected, because the result is an n-tuple that does not belong to the code. Regarding error correction it is important to note that after detecting the occurrence of these, we must decide which codeword is more likely to have been transmitted. Assuming that the codewords are equiprobable, we decide for the codeword nearest (in terms of Hamming distance) to the received n-tuple. Obviously this decision will be correct as long as an error pattern containing up to t errors per codeword occurs, satisfying the relation $2t + 1 \leq d$.

2.4   Error syndrome and decoding

Suppose a codeword $\mathbf{v}$ of a linear code with generator matrix $\mathbf{G}$ and parity-check matrix $\mathbf{H}$ is transmitted through a noisy channel. The signal associated with $\mathbf{v}$ arriving at the receiver is processed to produce an n-tuple $\mathbf{r}$ defined over the code alphabet. The n-tuple $\mathbf{r}$ may differ from $\mathbf{v}$ due to the noise added during transmission. The task of the decoder is to recover $\mathbf{v}$ from $\mathbf{r}$. The first step in decoding is to check whether $\mathbf{r}$ is a codeword. This process can be represented by the following expression:

$$\mathbf{r}\mathbf{H}^T = \mathbf{s},$$

where $\mathbf{s}$ denotes a vector with $n - k$ components, called *syndrome*. If $\mathbf{s} = \mathbf{0}$, i.e., a vector having the $n - k$ components equal to zero, we assume that no errors occurred, and thus $\mathbf{r} = \mathbf{v}$. However if $\mathbf{s} \neq \mathbf{0}$, $\mathbf{r}$ does not match a codeword in the row space of $\mathbf{G}$, and the decoder uses this error syndrome for detection, or for detection and correction. The received n-tuple $\mathbf{r}$ can be written as

$$\mathbf{r} = \mathbf{v} + \mathbf{e},$$

where $+$ denotes componentwise addition and $\mathbf{e}$ is defined over the code alphabet, denoting an n-tuple representing the error pattern.

The decoding process involves a decision about which codeword was transmitted. Considering a binary code, a systematic way to implement the decision process is to distribute the 2^n n-tuples into 2^k disjoint sets, each set having cardinality 2^{n-k}, so that each one of them contains only one codeword. Thus the decoding is done correctly if the received n-tuple $\mathbf{r}$ is in the subset of the transmitted codeword. We now describe one way of doing this. The 2^n binary n-tuples are separated into cosets as follows. The 2^k codewords are written in one row then, below the all-zero codeword, put an n-tuple $\mathbf{e}_1$ which is not present in the first row. Form the second row by adding modulo-2 to $\mathbf{e}_1$ the elements of the first row, as illustrated next.

$$
\begin{array}{ccccc}
0 & v_1 & v_2 & \cdots & v_{2^k-1} \\
e_1 & e_1 \oplus v_1 & e_1 \oplus v_2 & \cdots & e_1 \oplus v_{2^k-1},
\end{array}
$$

where $\oplus$ denotes modulo-2 addition of corresponding coordinates. Subsequent rows are formed similarly, and each new row begins with an element not previously used. In this manner, we obtain the array in Table 2.1, which is called *standard array*. The standard array rows are called *cosets* and the leftmost element in each coset is called a *coset leader*. The procedure used to construct the given linear code standard

Table 2.1. Standard array decomposition of an n-dimensional vector space over GF(2) using a block length n binary linear code having 2^k codewords.

0	v_1	v_2	$\cdots$	v_{2^k-1}
e_1	$e_1 \oplus v_1$	$e_1 \oplus v_2$	$\cdots$	$e_1 \oplus v_{2^k-1}$
e_2	$e_2 \oplus v_2$	$e_2 \oplus v_2$	$\cdots$	$e_2 \oplus v_{2^k-1}$
$\vdots$	$\vdots$	$\vdots$	$\cdots$	$\vdots$
$e_{2^{n-k}-1}$	$e_{2^{n-k}-1} \oplus v_1$	$e_{2^{n-k}-1} \oplus v_2$	$\cdots$	$e_{2^{n-k}-1} \oplus v_{2^k-1}$

array is called the *coset decomposition* of the vector space of n-tuples over GF(q).

To use the standard array it is necessary to find the row, and therefore, the associated leader, to which the incoming n-tuple belongs. This is usually not easy to implement because 2^{n-k} can be large, so that the concept of the standard array is most useful as a way to understand the structure of linear codes, rather than a practical decoding algorithm. Methods potentially practical for decoding linear codes are presented next.

2.4.1 Maximum likelihood decoding

If the codewords of a (n, k, d) code are selected independently and all have the same probability of being sent through a channel, an optimum way (in a sense we will explain shortly) to decode them is as follows. On receiving an n-tuple $\mathbf{r}$, the decoder compares it with all possible codewords. In the binary case, this means comparing $\mathbf{r}$ with 2^k distinct n-tuples that make up the code. The codeword nearest to $\mathbf{r}$ in terms of the Hamming distance is selected, i.e., we choose the codeword that differs from $\mathbf{r}$ in the least number of positions. This chosen codeword is supposedly the transmitted codeword. Unfortunately, the time necessary to decode a received n-tuple may become prohibitively long even for moderate values of k. It should be noted that the decoder must compare $\mathbf{r}$ with 2^k codewords, for a time interval corresponding to the duration of n channel digits. This fact makes this process of decoding inappropriate in many practical cases. A similar conclusion holds if one chooses to trade search time by a parallel decoder implementation, due to high decoder complexity.

Let $\mathbf{v}$ denote a codeword and let $P(\mathbf{r}|\mathbf{v})$ denote the probability of $\mathbf{r}$ being received when $\mathbf{v}$ is the transmitted codeword. If all codewords have the same probability of being transmitted then the probability $P(\mathbf{v}, \mathbf{r})$ of the pair $(\mathbf{v}, \mathbf{r})$ occurring is maximized when we select that $\mathbf{v}$ which maximizes $P(\mathbf{r}|\mathbf{v})$, known in statistics as the *likelihood function*.

2.4.2   Decoding by systematic search

A general procedure for decoding linear block codes consists of associating each nonzero syndrome with one correctable error pattern. One of the properties of the standard array is that all n-tuples belonging to the same coset have the same syndrome. Furthermore, each coset leader should be chosen as the most likely error pattern in the respective coset. Based in these standard array properties, it is possible to apply the following procedure for decoding:

(1) Calculate the syndrome for the received n-tuple.

(2) By systematic search, find the pattern of correctable errors, i.e., the coset leader, associated with the syndrome of the received n-tuple.

(3) Subtract from the received n-tuple the error pattern found in the previous step, to perform error-correction.

To implement this procedure it is necessary to generate successively all configurations of correctable errors and feed them into a combinational circuit, which gives as output the corresponding syndromes. Using a logic gate with multiple entries, we can detect when the locally generated syndrome coincides with the syndrome of the received n-tuple. If this (n, k, d) code corrects t errors per block then the number of distinct configurations of correctable errors that are necessary to generate by this systematic search process is given by

$$C_n^1 + C_n^2 + \cdots + C_n^t = \sum_{i=1}^{t} C_n^i \leq 2^{n-k} - 1. \tag{2.2}$$

It is easy to observe in (2.2) that the number of distinct configurations grows rapidly with n and t. For this reason, this decoding technique is of limited applicability.

2.4.3   Probabilistic decoding

In recent years, various decoding algorithms of a probabilistic nature, which in principle can operate on unquantized coordinate values of the received n-tuple have appeared in the literature. For practical reasons channel output quantization is employed. If the code used is binary and the number of channel output quantization levels is 2 then the decoding technique is called *hard-decision*, otherwise it is called a *soft-decision* decoding technique. A probabilistic decoding algorithm was introduced (Hartmann and Rudolph 1976) which is optimal in the sense that it minimizes the probability of error per digit, when the codewords are equiprobable and are transmitted in the presence of additive noise in a memoryless channel. This algorithm is exhaustive such that every codeword of the dual code is used in the decoding process. This feature makes it practical for use with high rate codes, contrary to what happens with most conventional techniques. Another decoding algorithm was introduced (Wolf 1978) which is a rule to walk in a trellis-type structure, and depends on the code **H** matrix. The received n-tuple is used to determine the most probable path in the trellis, i.e., the transmitted codeword. Trellis decoders for block codes for practical applications are addressed in (Honary and Markarian 1998).

2.5 Simple codes

In this section, we present some codes of relatively simple structure, which will allow the readers to understand more sophisticated coding mechanisms in future.

2.5.1 Repetition codes

A repetition code is characterized by the following parameters: $k = 1$, $n - k = c \geq 1$ and $n = k + c = 1 + c$. Because $k = 1$, this code has only two codewords, one is a sequence of n zeros and the other is a sequence of n 1's. The parity-check digits are all identical and are a repetition of the information digit. A simple decoding rule in this case is to declare the information digit transmitted as the one that most often occurs in the received word. This will always be possible when n is odd. If n is even, and there is a tie in the count of occurrence of 0's and 1's, we simply detect the occurrence of errors. The minimum distance of these codes is $d = n$ and their efficiency (or code rate) is $R = 1/n$. Obviously any pattern with $t \leq |n/2|$ errors is correctable.

2.5.2 Single parity-check codes

As the heading indicates, these codes have a single redundant digit per codeword. This redundant digit is calculated so as to make the number of 1's in the codeword even. That is, we count the number of 1's in the information section and if the result is odd the parity-check digit is made equal to 1, otherwise it is made equal to 0. The parameters of these codes are: $k \geq 1$, $n - k = 1$, i.e., $n = k + 1$. The Hamming distance and efficiency of these codes are, respectively, $d = 2$ and $R = k/n = k/(k+1)$.

The rule used for decoding single parity-check codes is to count the number of 1's in the received word. If the resulting count is even, the block received is assumed to be free of errors and is delivered to the recipient. Otherwise, the received block contains errors and the recipient is then notified of the fact. These codes, while allowing only to detect an odd number of errors, are effective when used in systems that operate with a return channel to request retransmission of messages, or when decoded with soft-decision.

2.5.3   Hamming codes

Hamming codes were the first nontrivial codes proposed for correcting errors (Hamming 1950). These codes are linear and have a minimum distance equal to 3, i.e., are capable of correcting one error per codeword. They have block length $n \leq 2^{n-k} - 1$, where $n - k$ is the number of redundant digits. This condition on n ensures the availability of sufficient redundancy to verify the occurrence of an error in a codeword, because the number of nonzero syndromes, $2^{n-k} - 1$, is always greater than or equal to the number of positions where an error can be.

EXAMPLE 2.5 *We next consider the construction of the code* $(7, 4, 3)$ *Hamming code. However, the ideas described here are easily generalized to any* $(n, k, 3)$ *Hamming code. The number of parity-check digits of the* $(7, 4, 3)$ *code is* $n - k = 7 - 4 = 3$. *Consider now the non-null binary numbers that can be written with* $n - k = 3$ *binary digits. That is,*

$$
\begin{array}{ccc c}
0 & 0 & 1 & c_1 \\
0 & 1 & 0 & c_2 \\
0 & 1 & 1 & k_1 \\
1 & 0 & 0 & c_3 \\
1 & 0 & 1 & k_2 \\
1 & 1 & 0 & k_3 \\
1 & 1 & 1 & k_4.
\end{array}
$$

Hamming associated the numbers of the form 2^j, $j = 0, 1, 2, \ldots$ *with parity-check positions. Other positions were associated with information digits, as indicated in the given list. Now, looking downward at the columns in the list, the parity-check equations denoted as* $c_i, 1 \leq i \leq 3$, *are written as modulo-2 sums of information positions where a 1 appears in the particular column considered. That is,*

$$
c_1 = k_1 \oplus k_2 \oplus k_4
$$
$$
c_2 = k_1 \oplus k_3 \oplus k_4
$$
$$
c_3 = k_2 \oplus k_3 \oplus k_4.
$$

Upon receiving a word, the decoder recalculates the parity-check digits and adds them modulo-2 to their corresponding parity-check digits in the received word to obtain the syndrome. If, for example, an error has hit *the digit k_3 the syndrome digits in positions c_2 and c_3 will be 1 and will indicate failure, while in position c_1 no failure is indicated because c_1 does not check k_3. The situation is represented as:*

$$(c_3, c_2, c_1) = (1, 1, 0),$$

which corresponds to the row for k_3 on the list considered. The error has thus been located and can then be corrected. Obviously, this procedure can be applied to any value of n.

Hamming codes are special in the sense that no other class of nontrivial codes can be so easily decoded and also because they are perfect, as defined next.

DEFINITION 2.6 *An (n, k, d) error-correcting code over* GF(q)*, which corrects t errors, is defined as perfect if and only if*

$$\sum_{i=0}^{t}(q-1)^i C_n^i = q^{n-k}.$$

With the exception of Hamming codes, the binary $(23, 12, 7)$ Golay code and the $(11, 6, 5)$ ternary Golay code, there are no other nontrivial linear perfect codes (Pless 1982, p.20). Nonlinear single error-correcting codes, with parameters identical to Hamming codes, were introduced in (Vasil'ev 1962).

2.6 Low-density parity-check codes

In 1993, the coding community was surprised with the discovery of turbo codes (Berrou, Glavieux, and Thitimajshima 1993), more than 40 years after Shannon's capacity theorem (Shannon 1948), referred to by many as Shannon's *promised land*. Turbo codes were the first capacity approaching practical codes. Not long after the discovery of turbo codes, their strongest competitors called low-density parity-check (LDPC) codes were rediscovered (MacKay and Neal 1996). LDPC codes have proved to perform better than turbo codes in many applications.

LDPC codes are linear block codes discovered by Gallager in 1960 (Gallager 1963), which have a decoding complexity that increases linearly with the block length. At the time of their discovery there was neither computational means for their implementation in practice nor to perform computer simulations. Some 20 years later a graphical representation of LDPC codes was introduced (Tanner 1981) which paved the way to their rediscovery, accompanied by further theoretical advances. It was shown that long LDPC codes with iterative decoding achieve a performance, in terms of error rate, very close to the Shannon capacity (MacKay and Neal 1996). LDPC codes have the following advantages with respect to turbo codes.

- LDPC codes do not require a long interleaver in order to achieve low error rates.

- LDPC codes achieve lower block error rates and their error floor occurs at lower bit error rates, for a decoder complexity comparable to that of turbo codes.

LDPC codes are defined by their parity-check matrix $\mathbf{H}$. Let ρ and γ denote positive integers, where ρ is small in comparison with the code block length and γ is small in comparison with the number of rows in $\mathbf{H}$.

DEFINITION 2.7 *A binary LDPC code is defined as the set of codewords that satisfy a parity-check matrix* $\mathbf{H}$, *where* $\mathbf{H}$ *has* ρ *1's per row and* γ *1's per column. The number of 1's in common between any two columns in* $\mathbf{H}$, *denoted by* λ, *is at most 1, i.e.,* $\lambda \leq 1$.

After their rediscovery by MacKay and Neal (1996) a number of good LDPC codes were constructed by computer search, which meant that such codes lacked in mathematical structure and consequently had more complex encoding than naturally systematic LDPC codes. The construction of systematic algebraic LDPC codes based on finite geometries was introduced in (Kou, Lin, and Fossorier 2001).

2.7   Problems with solutions

(1) Consider the vectors $\mathbf{v}_1 = (0, 1, 0, 0, 2)$ and $\mathbf{v}_2 = (1, 1, 0, 3, 2)$. Compute their respective Hamming weights, $W_H(\mathbf{v}_1)$ and $W_H(\mathbf{v}_2)$, and the Hamming distance $d_H(\mathbf{v}_1, \mathbf{v}_2)$.

Solution: The Hamming weights of $\mathbf{v}_1$ and $\mathbf{v}_2$ are, respectively, $W_H(\mathbf{v}_1) = 2$ and $W_H(\mathbf{v}_2) = 4$ and the Hamming distance between $\mathbf{v}_1$ and $\mathbf{v}_2$ is $d_H(\mathbf{v}_1, \mathbf{v}_2) = 2$.

(2) If d is an odd number, show that by adding an overall parity-check digit to the codewords of a binary (n, k, d) code, a $(n + 1, k, d + 1)$ code results.

Solution: The minimum nonzero weight of a linear code is equal to d, which in this problem is an odd number. Extending this binary code by appending an overall parity-check digit to each codeword will make the weight of every codeword an even number and thus the minimum nonzero weight will become $d + 1$. Therefore, the minimum distance of the extended code is $d + 1$.

(3) Consider the $(7, 3, 4)$ binary linear code having the following expressions for computing the redundant digits, also called parity-check digits.

$$c_1 = k_1 \oplus k_2, \quad c_2 = k_2 \oplus k_3,$$
$$c_3 = k_1 \oplus k_3, \quad c_4 = k_1 \oplus k_2 \oplus k_3.$$

Each block containing three information digits is encoded into a seven digit codeword. Determine the set of codewords for this code.

Solution: Employing the given parity-check equations, the following set of codewords results:

MESSAGES

$$\begin{bmatrix} 0 & 0 & 0 \\ 0 & 0 & 1 \\ 0 & 1 & 0 \\ 0 & 1 & 1 \\ 1 & 0 & 0 \\ 1 & 0 & 1 \\ 1 & 1 & 0 \\ 1 & 1 & 1 \\ k_1 & k_2 & k_3 \end{bmatrix} \Rightarrow$$

CODEWORDS

$$\begin{bmatrix} 0 & 0 & 0 & 0 & 0 & 0 & 0 \\ 0 & 0 & 1 & 0 & 1 & 1 & 1 \\ 0 & 1 & 0 & 1 & 1 & 0 & 1 \\ 0 & 1 & 1 & 1 & 0 & 1 & 0 \\ 1 & 0 & 0 & 1 & 0 & 1 & 1 \\ 1 & 0 & 1 & 1 & 1 & 0 & 0 \\ 1 & 1 & 0 & 0 & 1 & 1 & 0 \\ 1 & 1 & 1 & 0 & 0 & 0 & 1 \\ k_1 & k_2 & k_3 & c_1 & c_2 & c_3 & c_4 \end{bmatrix}$$

(4) Write the generator matrix and the parity-check matrix for the code in Problem 3, both in reduced echelon form.

Solution:

$$\mathbf{G} = \begin{bmatrix} 1 & 0 & 0 & 1 & 0 & 1 & 1 \\ 0 & 1 & 0 & 1 & 1 & 0 & 1 \\ 0 & 0 & 1 & 0 & 1 & 1 & 1 \\ k_1 & k_2 & k_3 & c_1 & c_2 & c_3 & c_4 \end{bmatrix}.$$

Therefore, it follows that

$$\mathbf{g} = \begin{bmatrix} 1 & 0 & 1 & 1 \\ 1 & 1 & 0 & 1 \\ 0 & 1 & 1 & 1 \end{bmatrix}, \quad \mathbf{h} = \mathbf{g}^T = \begin{bmatrix} 1 & 1 & 0 \\ 0 & 1 & 1 \\ 1 & 0 & 1 \\ 1 & 1 & 1 \end{bmatrix}.$$

The code parity-check matrix $\mathbf{H}$ takes the following form:

$$\mathbf{H} = \begin{bmatrix} 1 & 1 & 0 & 1 & 0 & 0 & 0 \\ 0 & 1 & 1 & 0 & 1 & 0 & 0 \\ 1 & 0 & 1 & 0 & 0 & 1 & 0 \\ 1 & 1 & 1 & 0 & 0 & 0 & 1 \\ k_1 & k_2 & k_3 & c_1 & c_2 & c_3 & c_4 \end{bmatrix}.$$

Chapter 3

CYCLIC CODES

Among the codes in the class of block codes cyclic codes are the most important from the point of view of practical engineering applications (Clark and Cain 1981, p.333). Cyclic codes are used in communication protocols (A, Györfi, and Massey 1992), in music CDs, in magnetic recording (Immink 1994), etc. This is due to their structure being based on discrete mathematics, which allows a considerable simplification in the implementation of encoders and decoders. The formal treatment of cyclic codes is done in terms of polynomial rings, with polynomial coefficients belonging to a Galois field GF(q), modulo $x^n - 1$, where n denotes the block length (Berlekamp 1968, p.119). However, a simple way to define cyclic codes is as follows.

DEFINITION 3.1 *A block code is called a cyclic code whenever a cyclic shift, applied to any of its codewords, produces a codeword in the same code, i.e., if* $\mathbf{v} = (v_0, v_1, v_2, \ldots, v_{n-1})$ *is a codeword then*

$$\mathbf{v}^i = (v_{n-i}, v_{n-i+1}, \ldots, v_0, v_1, \ldots, v_{n-i-1})$$

obtained by shifting $\mathbf{v}$ *cyclically by i places to the right, is also a codeword in the same code, considering the indices in v reduced modulo n.*

An n-tuple $\mathbf{v}$ can be represented by a polynomial of degree at most $n-1$ as follows:

$$v(x) = v_0 + v_1 x + v_2 x^2 + \cdots + v_{n-1} x^{n-1}.$$

Using properties of finite fields it can be shown that all the codewords of a (n, k, d) cyclic code are multiples of a well-defined polynomial $g(x)$, of degree $n - k$, and conversely that all polynomials of degree at most $n - 1$ which are divisible by $g(x)$ are codewords of this code (Lin and

Costello Jr. 2004, p.140). The polynomial $g(x)$ is called the code *generator polynomial* and is a factor of $x^n - 1$.

3.1   Matrix representation of a cyclic code

As we mentioned earlier, each codeword of a cyclic code is a multiple of the code generator polynomial $g(x)$. In this manner, it follows that the polynomials $g(x)$, $xg(x)$, $x^2g(x), \ldots, x^{k-1}g(x)$ are codewords. We also note that such codewords in particular are linearly independent, and thus can be used to construct a generator matrix $\mathbf{G}$ for the cyclic code which has $g(x)$ as its generator polynomial, as shown next.

$$\mathbf{G} = \begin{bmatrix} x^{k-1}g(x) \\ \vdots \\ x^2g(x) \\ xg(x) \\ g(x) \end{bmatrix},$$

where we assume that each row of $\mathbf{G}$ contains n elements, consisting of the coefficients of the corresponding row polynomial and the remaining empty positions are filled with zeros. For encoding purposes, the cyclic shift property of cyclic codes allows a sequential implementation of the $\mathbf{G}$ matrix which is presented next.

3.2   Encoder with $n - k$ shift-register stages

This encoding procedure is based on the property that each codeword in a cyclic code is a multiple of the code generator polynomial $g(x)$. The k information digits can be represented by a polynomial $I(x)$, of degree at most $k - 1$. Multiplying the polynomial $I(x)$ by x^{n-k} we obtain $I(x)x^{n-k}$, which is a polynomial of degree at most $n - 1$ which does not contain nonzero terms of degree lower than $n - k$. Dividing $I(x)x^{n-k}$ by $g(x)$ we obtain:

$$I(x)x^{n-k} = Q(x)g(x) + R(x),$$

where $Q(x)$ and $R(x)$ are, respectively, the quotient polynomial and the remainder polynomial. $R(x)$ has degree lower than $g(x)$, i.e., $R(x)$ has degree at most $n - k - 1$. If $R(x)$ is subtracted from $I(x)x^{n-k}$, the result is a multiple of $g(x)$, i.e., the result is a codeword. $R(x)$ represents the parity-check digits and has got no terms overlapping with $I(x)x^{n-k}$, as follows from our earlier considerations. The operations involved can be implemented with the circuit illustrated in Figure 3.1.

Let $g(x) = x^{n-k} + g_{n-k-1}x^{n-k-1} + \cdots + g_1x + 1$. The circuit in Figure 3.1 employs $n - k$ stages of a shift-register and pre-multiplies the

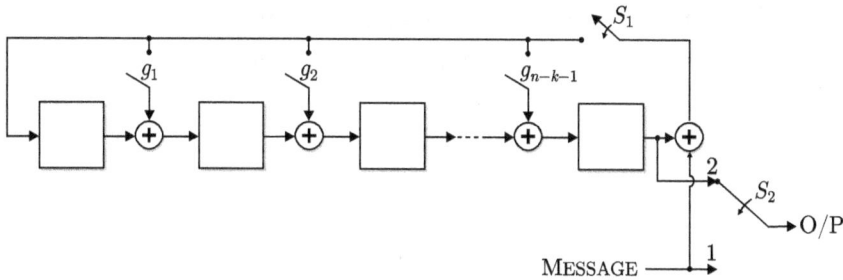

Figure 3.1. Encoder with $n - k$ shift-register stages for a binary cyclic code.

information polynomial $I(x)$ by x^{n-k}. A switch associated with the coefficient g_i, $i \in \{1, 2, \ldots, n - k - 1\}$, is closed if $g_i = 1$, otherwise it is left open. Initially the shift-register contents are 0s. Switch S_1 is closed and switch S_2 stays in position 1. The information digits are then simultaneously sent to the output and into the division circuit. After transmitting k information digits, the remainder, i.e., the parity-check digits, are the contents in the shift-register. Then, switch S_1 is open and switch S_2 is thrown to position 2. During the next $n - k$ clock pulses the parity-check digits are transmitted. This procedure is repeated for all subsequent k-digit information blocks. Another sequential encoding procedure exists for cyclic codes based on the polynomial $h(x) = (x^n - 1)/g(x)$, which employs k stages of shift-register. We chose not to present this procedure here, however, it can be easily found in the coding literature, for example, in the references (Clark and Cain 1981, p.73) and (Lin and Costello Jr. 2004, p.148). In the sequel we present a few classes of codes which benefit from the cyclic structure of their codewords.

3.3   Cyclic Hamming codes

The Hamming codes seen in Chapter 2 have a cyclic representation. Cyclic Hamming codes have a primitive polynomial $p(x)$ of degree m (Peterson and Weldon Jr. 1972, p.161) as their generator polynomial, and have the following parameters:

$$n = 2^m - 1, \ k = 2^m - m - 1, \ d = 3.$$

Cyclic Hamming codes are easily decodable by a Megitt decoder, or by an error-trapping decoder, which are described later. Due to the fact that Hamming codes are perfect codes (see Definition 2.6), very often they appear in the literature in most varied applications, as for example their codewords being used as protocol sequences for the collision channel without feedback (Rocha Jr. 1993).

3.4   Maximum-length-sequence codes

The maximum period possible for a q-ary sequence generated by a shift-register of m stages, employing linear feedback, is $q^m - 1$. We now look at binary maximum-length-sequence (m-sequence) codes, i.e., we consider $q = 2$. The m-sequence codes are cyclic, are dual codes of Hamming codes, and have the following parameters:

$$n = 2^m - 1, \ k = m, \ d = 2^{m-1}, \text{ for } m \geq 2.$$

The generator polynomial of an m-sequence code has the form $g(x) = (x^n - 1)/p(x)$, where $p(x)$ denotes a degree m primitive polynomial. The dictionary of an m-sequence code has an all-zeros codeword, and n nonzero codewords which result from n cyclic shifts of a nonzero codeword. It follows that all nonzero codewords have the same Hamming weight. The m-sequence codes are also called equidistant codes or simplex codes. The m-sequence codes are completely orthogonalizable in one-step (Massey 1963) and as a consequence they are easily decodable by majority logic. The nonzero codewords of an m-sequence code have many applications, including direct sequence spread spectrum, radar and location techniques.

3.5   Bose–Chaudhuri–Hocquenghem codes

Bose–Chaudhuri–Hocquenghem codes (BCH codes) were discovered independently and described in (Hocquenghem 1959) and (Bose and Ray–Chaudhuri 1960). The BCH codes are cyclic codes and represent one of the most important classes of block codes having algebraic decoding algorithms. For any two given positive integers m, t there is a BCH code with the following parameters:

$$n = q^m - 1, \ n - k \leq mt, \ d \geq 2t + 1.$$

The BCH codes can be seen as a generalization of Hamming codes, capable of correcting multiple errors in a codeword. One convenient manner of defining BCH codes is by specifying the roots of the generator polynomial.

DEFINITION 3.2 *A primitive BCH code over* GF(q), *capable of correcting* t *errors, having block length* $n = q^m - 1$, *has as roots of its generator polynomial* $g(x)$, α^{h_0}, α^{h_0+1}, ..., α^{h_0+2t-1}, *for any integer* h_0, *where* α *denotes a primitive element of* GF(q^m).

It follows that the generator polynomial $g(x)$ of a BCH code can be written as the least common multiple (LCM) of the minimal polynomials $m_i(x)$ (Berlekamp 1968, p.101), as explained next.

$$g(x) = \text{LCM}\{m_0(x), m_1(x), \ldots, m_{2t-1}(x)\},$$

where $m_i(x)$ denotes the minimal polynomial of α^{h_0+i}, $0 \le i \le 2t-1$. When α is not a primitive element $\mathrm{GF}(q^m)$ the resulting codes are called nonprimitive BCH codes. It follows that the respective block length is given by the multiplicative order of α. BCH codes with $h_0 = 1$ are called *narrow sense BCH codes*. An alternative definition for BCH codes can be given in terms of the finite field Fourier transform (see Appendix C) of the generator polynomial $g(x)$ (Blahut 1983, p.207). The roots α^{h_0+i}, $0 \le i \le 2t-1$, of $g(x)$ correspond to the zero components in the spectrum $G(z)$, in positions $h_0 + i$, $0 \le i \le 2t-1$.

DEFINITION 3.3 *A primitive BCH code over* $\mathrm{GF}(q)$*, capable of correcting* t *errors, having block length* $n = q^m - 1$*, is the set of all codewords over* $\mathrm{GF}(q)$ *whose spectrum is zero in* $2t$ *consecutive components* $h_0 + i$, $0 \le i \le 2t-1$.

The $2t$ consecutive roots of $g(x)$ or, equivalently, the $2t$ null spectral components of $G(z)$ guarantee a minimum distance $\delta = 2t + 1$, called the designed distance of the code, as shown next in a theorem known as the *BCH bound theorem*.

THEOREM 3.4 *Let* n *be a divisor of* $q^m - 1$*, for some positive integer* m. *If any nonzero vector* $\mathbf{v}$ *in* $\mathrm{GF}(q)^n$ *has a vector spectrum* $\mathbf{V}$ *with* $d-1$ *consecutive null components*, $V_j = 0$, $j = h_0, h_0+1, \ldots, h_0+d-2$, *then* $\mathbf{v}$ *has at least* d *nonzero components.*

Proof: Let us suppose by hypothesis that $\mathbf{v} = \{v_i\}$, $0 \le i \le n-1$, has a, $a < d$, nonzero components in positions $i_1, i_2, \ldots, i_a$ and that the finite field Fourier transform of $\mathbf{v}$ is identically zero in positions $h_0, h_0 + 1, \ldots, h_0 + d - 2$. We now define a frequency domain vector such that its inverse finite field Fourier transform has a zero whenever $v_i \neq 0$. One convenient choice for such a vector is based on the locator polynomial $L(z)$ as follows:

$$L(z) = \prod_{k=1}^{a}(1 - z\alpha^{i_k}) = L_a z^a + L_{a-1}z^{a-1} + \cdots + L_1 + L_0. \quad (3.1)$$

It follows that the spectral vector $\mathbf{L}$, associated with $L(z)$, is such that its inverse finite field Fourier transform $\mathbf{l} = \{l_i\}$, $0 \le i \le n-1$, has $l_i = 0$ precisely for all i such that $v_i \neq 0$, i.e., $l_i = 0$ whenever $v_i \neq 0$. It now follows that, in the time domain, we have $l_i v_i = 0$, $0 \le i \le n-1$, and consequently the corresponding finite field Fourier transform is all-zero:

$$\sum_{k=0}^{n-1} L_k V_{j-k} = \mathbf{0}, \quad (3.2)$$

i.e., the cyclic convolution in the frequency domain is zero. However $L(z)$ has degree at most $d - 1$ and $L_0 = 1$ in (3.1), thus (3.2) can be written as:

$$V_j = -\sum_{k=1}^{d-1} L_k V_{j-k}. \qquad (3.3)$$

Expression (3.3) shows how to generate the V_j's from a block of $d - 1$ known V_j's. However, by hypothesis, there is a block of $d - 1$ all-zero V_j's in positions $h_0, h_0 + 1, \ldots, h_0 + d - 2$. If such a block is used as an initial condition then all subsequent V_j's will be zero, and the vector **v** will be all-zero. Thus, any nonzero vector, with a spectrum having $d - 1$ consecutive null components, must necessarily have at least d nonzero components.   □

3.6   Reed–Solomon codes

Reed–Solomon (RS) codes (Blahut 1983, p.174) are nonbinary BCH codes, with $m = h_0 = 1$, defined by the parameters:

$$n = q - 1, \ n - k = 2t, \ d = 2t + 1.$$

Given an element α primitive in $\mathrm{GF}(q)$, the generator polynomial of an RS code has the following form:

$$g(x) = (x - \alpha)(x - \alpha^2)(x - \alpha^3) \cdots (x - \alpha^{2t}).$$

Many practical applications employ binary digits, therefore RS codes with $q = 2^r$ are a natural choice, having each 2^r-ary symbol represented by r binary digits. Since their minimum distance is equal to $n - k + 1$, RS codes constitute a class of *maximum distance separable* codes (MacWilliams and Sloane 1977, p.323). When mapped to binary, an RS code defined over $\mathrm{GF}(2^r)$ becomes a binary code of block length nr which is capable of correcting both random errors and burst errors, i.e., it can correct any combination of t erroneous binary r-tuples. Very often RS codes are employed as outer codes in serial concatenated coding schemes (Clark and Cain 1981, p.333).

3.7   Golay codes

The Golay codes have the following parameters: $n = 23$, $k = 12$, $d = 7$, with a binary alphabet, and $n = 11$, $k = 6$, $d = 5$, with a ternary alphabet. The Golay codes are the only existent perfect codes with $t > 1$ (Lint 1982, p.102).

3.7.1 The binary $(23, 12, 7)$ Golay code

The $(23, 12, 7)$ binary Golay code can be seen as a non-primitive BCH code generated as follows. Let α be a primitive element of $GF(2^{11})$ and notice that $2^{11} - 1 = 89 \times 23$. Thus, it follows that $\beta = \alpha^{89}$ is a nonprimitive element of order 23. The $(23, 12, 7)$ binary Golay code is specified by the generator polynomial $g(x)$ having β as a root. The roots of the minimal polynomial of β are:

$$\beta, \beta^2, \beta^4, \beta^8, \beta^{16}, \beta^9, \beta^{18}, \beta^{13}, \beta^3, \beta^6, \beta^{12},$$

where the fact that $\beta^{23} = 1$ has been used. The factoring of $x^{23} + 1$ over $GF(2)$ produces $x^{23} + 1 = (x+1)(x^{11} + x^{10} + x^6 + x^5 + x^4 + x^2 + 1)(x^{11} + x^9 + x^7 + x^6 + x^5 + x + 1)$. Depending on the choice of the primitive polynomial that generates $GF(2^{11})$, one of the two degree 11 factors of $x^{23} + 1$ will have β, β^2, β^3 and β^4 as the longest string of consecutive roots while the other degree 11 factor will have β^{19}, β^{20}, β^{21} and β^{22} as the longest string of consecutive roots. Any of these two degree 11 factors of $x^{23} + 1$ can be the code generator polynomial. We notice, however, that by the BCH bound (Lin and Costello Jr. 2004, p.205) this code has a designed distance $\delta = 5$, while the code true minimum distance is 7. The extended $(24, 12, 8)$ binary Golay code is obtained from the $(23, 12, 7)$ Golay code by appending to each codeword an overall parity-check digit, i.e., by appending a digit that makes even the total number of 1's in a codeword (see Chapter 2, Problem 2).

3.7.2 The ternary $(11, 6, 5)$ Golay code

The $(11, 6, 5)$ ternary Golay code can also be seen as a nonprimitive BCH code over $GF(3)$. It follows that if β is a primitive element in $GF(3^5)$ then $(\beta^{22})^{11} = \beta^{242} = 1$. Therefore, by considering $\alpha - \beta^{22}$, it follows that the powers $\alpha^i = (\beta^{22})^i$, $0 \leq i \leq 10$, are the roots of $x^{11} - 1$. The roots of $x^{11} - 1$ split into cyclotomic classes as $\{1\}$, $\{\alpha, \alpha^3, \alpha^9, \alpha^5, \alpha^4\}$, $\{\alpha^2, \alpha^6, \alpha^7, \alpha^{10}, \alpha^8\}$. Furthermore, $x^{11} - 1$ factors into irreducible polynomials over $GF(3)$ as

$$x^{11} - 1 = (x - 1)(x^5 + x^4 + 2x^3 + x^2 + 2)(x^5 + 2x^3 + x^2 + 2x + 2),$$

where the roots of $x^5 + x^4 + 2x^3 + x^2 + 2$ are $\{\alpha, \alpha^3, \alpha^9, \alpha^5, \alpha^4\}$ and the roots of $x^5 + 2x^3 + x^2 + 2x + 2$ are $\{\alpha^2, \alpha^6, \alpha^7, \alpha^{10}, \alpha^8\}$.

Let us consider the ternary cyclic $(11, 6, 5)$ Golay code with generator polynomial $g(x) = x^5 + x^4 + 2x^3 + x^2 + 2$. The set of roots of $g(x)$ contains a maximum of three consecutive roots, namely α^3, α^4 and α^5, and thus by the BCH bound it guarantees a minimum Hamming distance of 4 but not 5 as required for a double-error correcting code. We can also consider the $(11, 6, 5)$ ternary Golay code generated by $x^5 + 2x^3 + x^2 + 2x + 2$, having α^6, α^7 and α^8 as its longest string of consecutive roots of the generator polynomial. The extended $(12, 6, 6)$ ternary Golay code is obtained from the $(11, 6, 5)$ Golay code by appending to each codeword an overall parity-check digit, i.e., by appending a digit that makes equal to zero the modulo 3 sum of the digits in a codeword.

3.8   Reed-Muller codes

Reed–Muller (RM) codes are binary codes which are equivalent to cyclic codes with an overall parity-check digit attached. Prior to defining RM codes, it is necessary to define the Hadamard product of vectors of the same length, as the vector whose components are the product of the corresponding components of the factors. For example, the **abc** Hadamard product of vectors $\mathbf{a} = (a_1, a_2, \ldots, a_n)$, $\mathbf{b} = (b_1, b_2, \ldots, b_n)$ and $\mathbf{c} = (c_1, c_2, \ldots, c_n)$ is given by:

$$\mathbf{abc} = (a_1 b_1 c_1, a_2 b_2 c_2, \ldots, a_n b_n c_n).$$

DEFINITION 3.5 *Let $\mathbf{v_0}$ be the vector having 2^m components, all of which are equal to 1. Let $\mathbf{v_1}, \mathbf{v_2}, \ldots, \mathbf{v_m}$ be vectors forming the rows of a matrix whose 2^m columns are all the distinct binary m-tuples. The RM code of order r is defined by the generator matrix whose rows are the vectors $\mathbf{v_0}, \mathbf{v_1}, \ldots, \mathbf{v_m}$ and their respective Hadamard products two at a time, three at a time, ..., r at a time. For any positive integer m, the RM code of order r has the following parameters:*

$$n = 2^m, \quad k = \sum_{i=0}^{r} C_m^i, \quad d = 2^{m-r}.$$

The RM codes are a large class of codes but have an error correction power in general lower than that of BCH codes of equivalent rate. Two important aspects characterize RM codes. The first is the fact that they are easily decodable by majority logic and the second is to be a sub-class of codes constructed from the Euclidean geometry. We will revisit RM codes with more in-depth detail in Section 11.5.

3.9   Quadratic residue codes

Before defining quadratic residue codes it is necessary to introduce *quadratic residues*.

DEFINITION 3.6 *An integer r is a quadratic residue of a prime number p, if and only if there exists an integer s such that $s^2 \equiv r \mod p$, where the abbreviation mod is used to mean modulo.*

It can be shown (MacWilliams and Sloane 1977, p.481) that if $n = 8m \pm 1$ is a prime number then 2 is a quadratic residue of n. In this case $x^n + 1$ can be factored as $(x + 1)G_r(x)G_{\bar{r}}(x)$, where:

$$G_r(x) = \prod_{r \in R_0} (x + \alpha^r) \quad \text{and} \quad G_{\bar{r}}(x) = \prod_{\bar{r} \in \overline{R_0}} (x + \alpha^{\bar{r}}),$$

where α denotes an element of multiplicative order n in an extension field of GF(2), R_0 denotes the set of quadratic residues modulo n and $\overline{R_0}$ denotes the set of quadratic non-residues modulo n. The cyclic codes with generator polynomials $G_r(x)$, $(1+x)G_r(x)$, $G_{\bar{r}}(x)$ and $(x+1)G_{\bar{r}}(x)$ are called *quadratic residue codes*. An efficient way of decoding quadratic residue codes is by means of permutations (p.513). Permutation decoders are usually more complex than algebraic BCH decoders. However, the minimum distance of some quadratic residue codes of moderate length is greater than the minimum distance of BCH codes of comparable block length.

3.10   Alternant codes

The class of alternant codes (p.332) is of great importance for including, as particular cases, the BCH codes, Goppa codes, Srivastava codes and Chien-Choy codes. Alternant codes are constructed by a simple modification of the parity-check matrix of a BCH code. A BCH code of block length n and designed distance δ over GF(q) has a parity-check matrix $\mathbf{H} = [h_{ij}]$, where $h_{ij} = \alpha^{ij}$, $1 \le i \le \delta - 1, 0 \le j \le n - 1$, and $\alpha \in \mathrm{GF}(q^m)$ is a primitive nth root of unity (p.196). Alternant codes are obtained by considering $h_{ij} = x_j^{i-1}y_j$, where $\mathbf{x} = (x_1, x_2, \ldots, x_n)$ is a vector with distinct components, belonging to GF(q^m) and $\mathbf{y} = (y_1, y_2, \ldots, y_n)$ is a vector with nonzero components, also belonging to GF(q^m). For more detailed information on this subject the reader should consult Chapter 12 of (MacWilliams and Sloane 1977).

3.11   Problems with solutions

(1) Consider the $(2^m - 1, 2^m - m - 2, 3)$ binary cyclic Hamming code generated by $p(x)$, a primitive polynomial of degree m. Show that this code is capable of correcting any pattern containing at most two erasures.

Solution: By changing the values of at most two positions in a codeword of a code with minimum distance 3 cannot produce another codeword. Therefore by trial and error, at most four binary

Table 3.1. Galois field GF(8) generated by $p(x) = x^3 + x + 1$.

Exponential form	Polynomial form
0	0
1	1
α	α
α^2	α^2
α^3	$\alpha + 1$
α^4	$\alpha^2 + \alpha$
α^5	$\alpha^2 + \alpha + 1$
α^6	$\alpha^2 + 1$

patterns need to be tested as replacements for two erasures, which are corrected in this manner.

(2) Construct a table with the elements of GF(8) expressed as powers of a primitive element α which is a root of $x^3 + x + 1$.

Solution: Table 3.1 presents the elements of GF(8) as powers of a primitive root of the polynomial $x^3 + x + 1$.

(3) Construct a table with the elements of GF(16) expressed as powers of a primitive element α which is a root of $x^4 + x + 1$.

Solution: Table 3.2 presents the elements of GF(16) as powers of a primitive root of the polynomial $x^4 + x + 1$.

Table 3.2. Galois field GF(16) generated by $p(x) = x^4 + x + 1$.

Exponential form	Polynomial form
0	0
1	1
α	α
α^2	α^2
α^3	α^3
α^4	$\alpha + 1$
α^5	$\alpha^2 + \alpha$
α^6	$\alpha^3 + \alpha^2$
α^7	$\alpha^3 + \alpha + 1$
α^8	$\alpha^2 + 1$
α^9	$\alpha^3 + \alpha$
α^{10}	$\alpha^2 + \alpha + 1$
α^{11}	$\alpha^3 + \alpha^2 + \alpha$
α^{12}	$\alpha^3 + \alpha^2 + \alpha + 1$
α^{13}	$\alpha^3 + \alpha^2 + 1$
α^{14}	$\alpha^3 + 1$

(4) Let α be a root in GF(8) of the binary primitive polynomial $p(x) = x^3 + x + 1$. Determine the generator polynomial $g(x)$ of an RS code in GF(8) with minimum distance 5, such that $g(\alpha) = 0$.

Solution: The generator polynomial $g(x)$ is required to have four consecutive roots, where α is one of them. Using the roots $\alpha, \alpha^2, \alpha^3$ and α^4, and using Table 3.1 to simplify the result we obtain $g(x) = x^4 + \alpha^3 x^3 + x^2 + \alpha x + \alpha^3$. Verify that $\alpha^4, \alpha^5, \alpha^6$ and α, although not all consecutive roots, offer another solution to this problem.

(5) Determine the generator polynomial $g(x)$ of the $(15, 5, 7)$ binary BCH code, having roots α^i, $1 \leq i \leq 6$.

Solution: Let $g(x) = m_1(x)m_3(x)m_5(x)$ denote the generator polynomial. Using Table 3.2 we obtain $m_1(x) = x^4 + x + 1$, $m_3(x) = x^4 + x^3 + x^2 + x + 1$, and $m_5(x) = x^2 + x + 1$. Simplifying, we obtain $g(x) = x^{10} + x^8 + x^5 + x^4 + x^2 + x + 1$.

Chapter 4

DECODING CYCLIC CODES

The decoding procedures for linear block codes are also applicable to cyclic codes. However, the algebraic properties associated with the cyclic structure allow important simplifications when implementing a decoder for a cyclic code, both for calculating the syndrome and for correcting errors. The syndrome computation consists in dividing by the generator polynomial $g(x)$ the polynomial representing the word received from the channel. The remainder of this division is the syndrome, denoted by $s(x)$. If $s(x) = 0$, the received word is accepted as being a codeword. Otherwise, i.e., if $s(x) \neq 0$, we declare that errors have occurred. In this manner, it is clear that a circuit to detect errors with a cyclic code is rather simple. The problem of locating error positions in a received word for correction is a different matter and in general requires more elaborate techniques to be implemented in practice. So far, the most important algebraic decoding techniques for cyclic codes are those based on the Berlekamp–Massey algorithm (Massey 1969) and on the Euclidean algorithm (MacWilliams and Sloane 1977, pp.362–365), (Clark and Cain 1981, pp.216–218). In the sequel, we give a brief presentation of the more relevant decoding procedures for cyclic codes.

4.1 Meggitt decoder

The Meggitt decoding algorithm (Lin and Costello Jr. 2004, p.156) consists in employing a circuit to identify those syndromes that correspond to error patterns containing an error in the highest order position of the received word, i.e., an error at position x^{n-1}. Thus, when this digit is being delivered to the sink it can be altered or not, depending on the decision determined by the circuit that identifies errors in that position. If we cyclically shift the received word, digit by digit with a shift-register, all the digits in this word will necessarily occupy position x^{n-1} after a certain number of shifts, and will be examined by the circuit that identifies errors in that position. The decision for choosing the Meggitt decoder depends on the complexity of the circuit that identifies errors at position x^{n-1} of the received word. A straightforward way of implementing a Meggitt decoder is by the use of programmable memories.

4.2   Error-trapping decoder

The decoding algorithm known as *error trapping* (Lin and Costello Jr. 2004, p.166) operates by cyclically shifting, bit by bit, the syndrome of the received word, and observing its Hamming weight at each step. If a syndrome of Hamming weight at most t (number of correctable errors) is detected, then the corresponding length $n-k$ segment of the error vector, containing all the errors, coincides with this shifted syndrome. Once the error pattern is identified, or trapped, error correction is immediate. However, a situation may occur where, after n cyclic shifts, the syndrome Hamming weight was never t or less. In this case the decoder declares the occurrence of an error pattern that spreads over a length of cyclically consecutive positions greater than $n - k$. This decoding procedure is more appropriate for use with low rate codes, since for a code with block length n, having k information digits and capable of correcting t errors per block, the efficient application of error trapping requires the condition

$$n/k = 1/R > t$$

to be satisfied.

4.3   Information set decoding

In a linear (n, k, d) code any set of k positions, that can be independently specified in a codeword, constitutes an *information set* (Clark and Cain 1981, p.102). As a consequence, the symbols in an information set define a codeword. If an information set in a received n-tuple contains no errors then it is possible to reconstitute the transmitted codeword. The decoding algorithm based on information sets consists of the following steps:

(1) Construct several information sets for the given code.

(2) Form various estimates of the transmitted codeword, by decoding the received word using each of the information sets obtained in the previous step.

(3) Compare the received word with the estimates obtained in the previous step and decide for the codeword nearest to the received word.

4.4 Threshold decoding

Threshold decoding (Massey 1963) is also known as *majority logic decoding*. A great number of applications of this technique is concentrated on cyclic codes. Before describing threshold decoding, it is necessary to introduce the concept of *parity-check sums* or simply *parity sums*. As we have seen earlier, the syndrome is represented by a vector having $n-k$ coordinates, $\mathbf{s} = (s_0, s_1, \ldots, s_{n-k-1})$, where, for a given (n, k, d) linear code with parity-check matrix $\mathbf{H} = [h_{ij}]$ and an error vector $(e_0, e_1, \ldots, e_{n-1})$ we have the relation:

$$\mathbf{s} = \mathbf{e}\mathbf{H}^T,$$

where each syndrome component s_j, $0 \leq j \leq n - k - 1$, is given by

$$s_j = \sum_{i=0}^{n-1} e_i h_{ij}. \tag{4.1}$$

The linear combination of syndrome digits

$$A = \sum_{i=0}^{n-k-1} a_i s_i,$$

where a_i is either 0 or 1, with the help of (4.1) can be written as

$$A = \sum_{i=0}^{n-1} b_i e_i, \tag{4.2}$$

where b_i is either 1 or 0. An error in position e_l is said to be checked by A if the corresponding coefficient b_i in (4.2) is 1. Expression (4.2) is called a *parity-check sum*.

DEFINITION 4.1 *Given a set of J parity-check sums $A_1, A_2, \ldots, A_J$, such that a position e_l in the error vector is checked by all of them and all other positions e_i, $i \neq l$, in the error vector are checked at most once, then this set is said to be orthogonal on position e_l.*

Assuming the occurrence of $t \leq \lfloor J/2 \rfloor$ errors in the received word, the threshold decoding procedure is based on the following reasoning:

(i) $e_l = 1$. This means that the remaining $t - 1$ errors will affect at most $t - 1 \leq \lfloor J/2 \rfloor - 1$ of the J parity sums, thus leaving at least $J - \lfloor J/2 \rfloor + 1 = \lceil J/2 \rceil + 1$ of the parity sums equal to 1.

(ii) $e_l = 0$. In this case the t errors will affect at most $\lfloor J/2 \rfloor$ parity sums, i.e., half of the parity sums in the worst case situation.

The threshold decoding rule consists in making $e_l = 1$, i.e., to declare the presence of an error in this position, whenever the majority of the parity sums are equal to 1. Otherwise, make $e_l = 0$, i.e., in case the number of parity sums equal to 1 coincides with the number of parity sums equal to 0, or if the majority of the parity sums are equal to 0. When the code considered is a cyclic code, after decoding a given position in a codeword a cyclic shift is applied to that codeword and the same set of J parity-check sums are used to decode position e_{l-1}, and so on in a similar manner for decoding the remaining codeword positions until the complete codeword is decoded.

Codes for which $J = d - 1$ are said to be completely orthogonalizable. However, it is not always possible to obtain directly J parity sums orthogonal on a given position of a codeword. For various classes of codes (Lin and Costello Jr. 2004, p.296) orthogonal parity sums are obtained in L steps, where at each step one uses parity sums orthogonal on a sum of codeword positions. This topic of threshold decoding is treated in greater detail in Chapter 11.

4.5   Algebraic decoding

In general, for a given (n, k, d) code the decoding process has always a complexity higher than the corresponding encoding process. Thus, from a practical point of view the best code is chosen subject to a specified budget. This financial constraint can force the choice of suboptimum codes, however having a decoder which is amenable to practical implementation. Fortunately, for some classes of algebraic codes decoding algorithms were developed which are computationally efficient (Berlekamp 1968, p.178). The problem of decoding algebraic codes consists in solving a system of nonlinear equations, whose direct solution in general is not obvious.

4.5.1   Berlekamp-Massey time domain decoding

An important algebraic decoding algorithm for BCH codes was published in (Berlekamp 1968). Analyzing Berlekamp's algorithm, Massey (1969) showed that it provided a general solution to the problem of synthesizing the shortest linear feedback shift register capable of generating a prescribed finite sequence of digits. Since then this algorithm is known as the Berlekamp–Massey (BM) algorithm. The BM algorithm is widely applicable for decoding algebraic codes, including RS codes and BCH codes. For binary BCH codes, there is no need to calculate

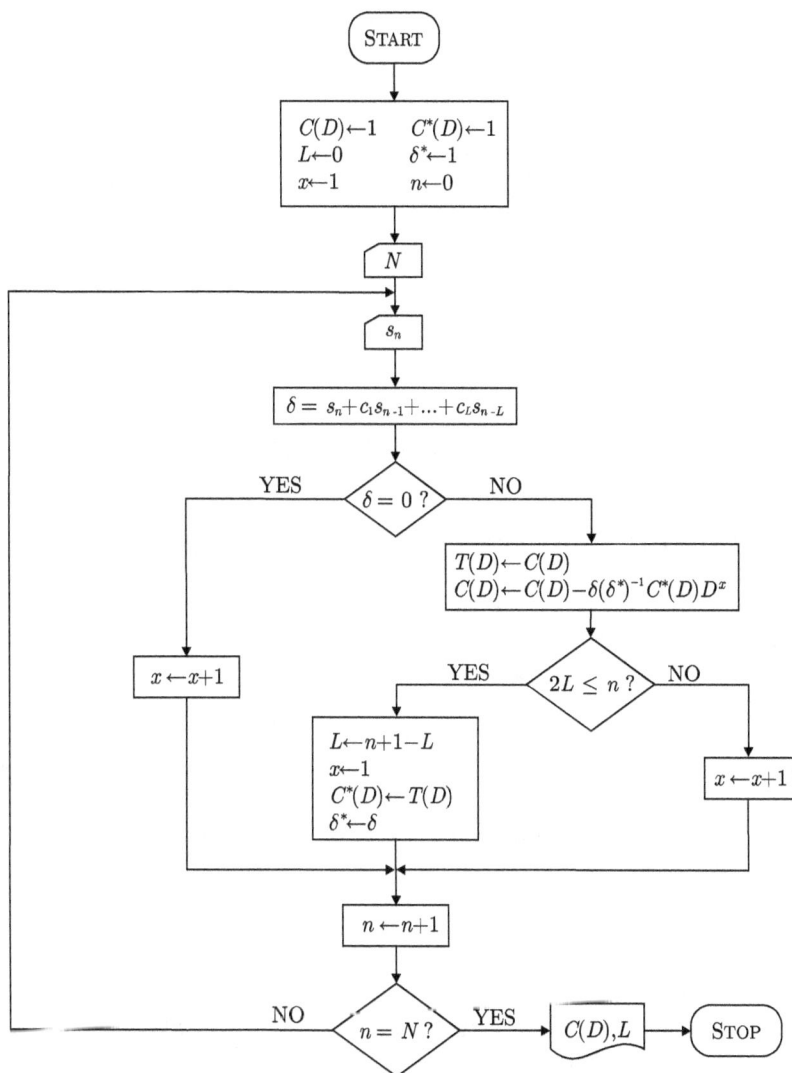

Figure 4.1. Flow chart of the Berlekamp–Massey time domain decoding algorithm.

error magnitude values, since in GF(2) it is sufficient to determine the positions of the errors to perform error correction. For non-binary BCH codes, including RS codes, both error location and error magnitudes have to be determined to perform error correction.

Let (n, k, d) denote an algebraic code (BCH or RS) with generator polynomial $g(x)$ having coefficients in some finite field GF(q) and having roots $\alpha, \alpha^2, \ldots, \alpha^{2t}$. Let $c(x) = \sum_{i=0}^{n-1} c_i x^i$ denote a codeword polynomial, let $e(x) = \sum_{i=0}^{n-1} e_i x^i$ denote the error polynomial with coefficients in GF(q) and let $r(x) = c(x) + e(x)$, with addition over GF(q), denote the received n-tuple in polynomial form. In general, the approach used for algebraic decoding contains the following steps.

(1) Compute the first $2t$ coefficients $s_0, s_1, \ldots, s_{2t-1}$ of the syndrome polynomial $s(x) = s_0 + s_1 x + \cdots + s_{2t-1} x^{2t-1} + \cdots$, where $s_0 = r(\alpha), s_1 = r(\alpha^2), \ldots, s_{2t-1} = r(\alpha^{2t})$.

(2) Use the sequence $s_0, s_1, \ldots, s_{2t-1}$ as input to the BM algorithm and compute the error-locator polynomial $\sigma(x)$, of degree τ, $\tau \leq t$, where

$$\sigma(x) = 1 + \sigma_1 x + \sigma_2 x^2 + \cdots + \sigma_\tau x^\tau.$$

(3) Find the roots of $\sigma(x)$, denoted by $\beta_1, \beta_2, \ldots, \beta_\tau$, whose multiplicative inverses give the error locations.

(4) Compute the error magnitudes in case of nonbinary codes.

In (Berlekamp 1968, p.220) a procedure was introduced for computing error magnitudes for nonbinary cyclic codes and defined the polynomial

$$\begin{aligned} Z(x) \;=\; & 1 + (s_0 + \sigma_1)x + (s_1 + \sigma_1 s_0 + \sigma_2)x^2 + \cdots \\ & \cdots + (s_{\tau-1} + \sigma_1 s_{\tau-2} + \sigma_2 s_{\tau-3} + \cdots + \sigma_\tau)x^\tau. \end{aligned} \quad (4.3)$$

Error magnitudes at positions β_i^{-1}, $1 \leq i \leq \tau$, are calculated as

$$e_i = \frac{Z(\beta_i)}{\prod_{j=1, j \neq i}^{\tau}(1 - \beta_j^{-1}\beta_i)}. \quad (4.4)$$

The block diagram in Figure 4.1 (Massey 1998) illustrates the steps for running the BM algorithm, where $C(D)$ plays the role of $\sigma(x)$. As a result of this procedure, an estimate of the error pattern that has minimum Hamming weight and satisfies all the syndrome equations is obtained. Whether such an estimate will be the true error pattern that occurred on the channel or not will depend on its Hamming weight τ in comparison to t. Decoding succeeds whenever the condition $\tau \leq t$ is satisfied.

4.5.2   Euclidean frequency domain decoding

The decoding procedure known as *algebraic decoding in the frequency domain* (Blahut 1983, p.193) is applicable to any cyclic code, being however more efficient for BCH codes. The received vector **r** is assumed to result from the sum **v**+**e** over GF(q) of a codeword **v** and an error vector

e, i.e., **r** = **v** + **e**. The steps required for using this decoding procedure are now described.

(1) **Fourier transform of the received vector**
   The Fourier transform **R** of the received vector **r** is computed. Due to linearity **R** can be expressed as a function of the corresponding Fourier transforms of **v** and **e**, i.e., as **R** = **V** + **E**.

(2) **Syndrome calculation**

$$S(z) = \sum_{j=0}^{2t-1} S_j z^j,$$

$$S_j = R_{j+h_0} = \sum_{i=0}^{n-1} r_i \alpha^{i(j+h_0)}, j = 0, 1, \ldots, 2t - 1.$$

(3) **Error locator polynomial calculation**
   This step is also known as the *solution of the key equation* (Clark and Cain 1981, p.189). The error locator polynomial, $L(z)$, has the form:

$$L(z) = \sum_{k=0}^{\nu-1} L_z z^k = \prod_{k=1}^{\nu} (1 - z\alpha^{i_k}),$$

where $\nu \leq t$ and $i_1, i_2, \ldots, i_\nu$, correspond to the error locations. The Euclidean algorithm (see Appendix C) is applied with $a(z) = z^{2t}$ and $b(z) = S(z)$, stopping when the degree of $r_i(z)$ becomes less than t. We take $L(z) = g_i(z)$.

(4) **Error vector Fourier transform calculation**
   The error vector **e** has its Fourier transform denoted by **E**, which has the following polynomial representation:

$$E(z) = \sum_{j=0}^{n-1} E_j z^j.$$

Since the code is assumed to have $2t$ consecutive roots, it follows that the coefficients $E_j, j = h_0, h_0 + 1, \ldots, h_0 + 2t - 1$, are known. The remaining unknown coefficients are calculated by means of the following recursive formula:

$$\sum_{k=0}^{n-1} L_k E_{j-k} = 0,$$

which can be slightly simplified to

$$E_j = \sum_{k=0}^{\nu-1} L_k E_{j-k},$$

because $L_0 = 1$ and $L(z)$ has degree $\nu - 1 \leq t - 1$.

(5) **Error correction**
The error vector is obtained as the inverse Fourier transform of vector $\mathbf{E}$, obtained in step 4. Finally, the estimated transmitted codeword is given by

$$\mathbf{v} = \mathbf{r} - \mathbf{e}.$$

4.6   Soft-decision decoding

Probabilistic decoding algorithms (see Section 2.4.3) are used in soft-decision decoding. A soft-decision decoder employs received symbol reliability information supplied by the demodulator when deciding which codeword was transmitted. Typically soft-decision decoding leads to a gain of at least 2.0 dB with respect to hard-decision decoding.

4.6.1   Decoding LDPC codes

Efficient decoding of LDPC codes relies on the sum-product algorithm (SPA) (Moreira and Farrell 2006, p.282), which is a symbol-by-symbol soft-in soft-out iterative decoding algorithm. The received symbols are fed to the decoder which employs iterations based on the code parity-check matrix $\mathbf{H}$ to improve the reliability of the decoded symbols. After each iteration the available symbol reliability values are used to make hard decisions and to output a decoded binary n-tuple $\mathbf{z}$. If $\mathbf{z}\mathbf{H}^{\mathrm{T}} = \mathbf{0}$ then $\mathbf{z}$ is a codeword and decoding stops. If $\mathbf{z}\mathbf{H}^{\mathrm{T}} \neq \mathbf{0}$ then $\mathbf{z}$ is not a codeword and another iteration is performed by the decoder. A stop condition is defined by specifying a maximum number of decoder iterations. A decoding failure occurs if the maximum number of iterations is reached and the decoder does not find a codeword.

Let $\mathbf{v} = (v_1, v_2, \ldots, v_i, \ldots, v_n)$ denote a codeword and let $\mathbf{y}$ denote a received n-tuple with real valued coordinates. The SPA is implemented by computing the marginal probabilities $P(v_i|\mathbf{y})$, for $1 \leq i \leq n$. A detailed description of SPA decoding of LDPC codes is available in (MacKay 1999).

4.7 Problems with solutions

(1) Consider the binary maximum-length sequence $(7, 3, 4)$ code generated by $g(x) = x^4 + x^3 + x^2 + 1$. Determine three parity-check sums, orthogonal on position e_6.

Solution: The following three parity-check sums, orthogonal on position e_6, are obtained:

$$\begin{aligned}
A_1 &= e_6 + e_4 + e_3 \\
A_2 &= e_6 + e_5 + e_1 \\
A_3 &= e_6 + e_2 + e_0.
\end{aligned}$$

Since $J = 3 = d - 1$, we conclude that this code is completely orthogonalizable in one-step, and by threshold decoding it corrects one error per codeword.

(2) Consider the binary cyclic $(7, 4, 3)$ Hamming code generated by $g(x) = x^3 + x + 1$. This code is the dual of the $(7, 3, 4)$ code of Problem 1 and is threshold decodable in two-steps. Describe a two-step decoding algorithm for this code, with the help of the code **H** matrix.

Solution: We now describe the solution with the help of the code **H** matrix.

$$\mathbf{H} = \begin{bmatrix} 1 & 1 & 1 & 0 & 1 & 0 & 0 \\ 0 & 1 & 1 & 1 & 0 & 1 & 0 \\ 1 & 1 & 0 & 1 & 0 & 0 & 1 \end{bmatrix}.$$

Using the **H** matrix we obtain the following parity sums:

$$\begin{aligned}
A_1 &= s_2 &= e_6 + e_5 + e_4 + e_2 \\
A_2 &= s_2 + s_1 &= e_6 + e_3 + e_2 + e_1 \\
A_3 &= s_0 &= e_6 + e_5 + e_3 + e_0.
\end{aligned}$$

Notice that A_1 and A_2 are orthogonal on the sum $e_6 + e_2$, and that A_1 and A_3 are orthogonal on the sum $e_6 + e_5$. We now use the sums A_1 and A_2 to estimate the sum $e_6 + e_2$, and use the sums A_1 and A_3 to estimate $e_6 + e_5$. This concludes the first decoding step. We now use the two estimated sums, denoted by A_4 and A_5, respectively, i.e.,

$$\begin{aligned}
A_4 &= e_6 + e_2 \\
A_5 &= e_6 + e_5,
\end{aligned}$$

which are orthogonal on e_6, constitute the second decoding step and complete the orthogonalization process for this code.

(3) Consider the $(15, 7, 5)$ binary BCH code, with $g(x) = m_1(x)m_3(x) = (x^4 + x + 1)(x^4 + x^3 + x^2 + x + 1)$ as generator polynomial and decode the received polynomial $r(x) = x + x^9$ with the Berlekamp–Massey algorithm.

Solution: The syndrome coefficients for the received polynomial $r(x) = x + x^9$ are calculated with the help of Table 3.2 as

$$
\begin{aligned}
s_0 &= r(\alpha)   &&= \alpha + \alpha^9   &&= \alpha^3 \\
s_1 &= r(\alpha^2) &&= \alpha^2 + \alpha^{18}   &&= \alpha^6 \\
s_2 &= r(\alpha^3) &&= \alpha^3 + \alpha^{27}   &&= \alpha^{10} \\
s_3 &= r(\alpha^4) &&= \alpha^4 + \alpha^{36}   &&= \alpha^{12}.
\end{aligned}
$$

Table 4.1 is used to compute the error-locator polynomial with the BM algorithm for the input sequence $s_0, s_1, s_2, s_3 = \alpha^3, \alpha^6, \alpha^{10}, \alpha^{12}$. The polynomial $C(D) = 1 + \alpha^3 D + \alpha^{10} D^2$ is the error-locator polynomial, and since $C(D) = \sigma(x)$ we write

$$\sigma(x) = 1 + \alpha^3 x + \alpha^{10} x^2,$$

the roots of which are found by an exhaustive search to be $\beta_1 = \alpha^6$ and $\beta_2 = \alpha^{14}$. The error positions are the exponents of α in the representation of β_1^{-1} and β_2^{-1}, i.e., $\beta_1^{-1} = \alpha^{-6} = \alpha^9$ and $\beta_2^{-1} = \alpha^{-14} = \alpha$. Two errors are thus located, at positions x and x^9, respectively. After the operation of error correction is completed, the decoded word is the all-zero codeword, i.e., $c(x) = 0$.

(4) Consider the $(7, 5, 3)$ RS code over GF(8), with generator polynomial $g(x) = (x - \alpha)(x - \alpha^2)$ and decode the received polynomial $r(x) = \alpha^2 x^3$ using the Berlekamp–Massey algorithm.

Table 4.1. Evolution of the BM algorithm for the input sequence $(s_0, s_1, s_2, s_3) = (\alpha^3, \alpha^6, \alpha^{10}, \alpha^{12})$.

n	s_n	δ	$T(D)$	$C(D) = \sigma(X)$	L	$C^*(D)$	δ^*	x
0	α^3	α^3	–	1	0	1	1	1
1	α^6	0	1	$1 + \alpha^3 D$	1	1	α^3	1
2	α^{10}	α^{13}	1	$1 + \alpha^3 D$	1	1	α^3	2
3	α^{12}	0	$1 + \alpha^3 D$	$1 + \alpha^3 D + \alpha^{10} D^2$	2	$1 + \alpha^3 D$	α^{13}	1

Table 4.2. Evolution of the BM algorithm for the input sequence $(s_0, s_1) = (\alpha^5, \alpha)$.

n	s_n	δ	$T(D)$	$C(D) = \sigma(X)$	L	$C^*(D)$	δ^*	x
0	α^5	α^5	—	1	0	1	1	1
1	α	1	1	$1 + \alpha^5 D$	1	1	α^5	1
			$1 + \alpha^5 D$	$1 + \alpha^3 D$	1	$1 + \alpha^5 D$	1	1

Solution: The syndrome coefficients for the received polynomial $r(x) = \alpha^2 x^3$ are calculated with the help of Table 3.1 as

$$
\begin{aligned}
s_0 &= r(\alpha) &= \alpha^5 \\
s_1 &= r(\alpha^2) &= \alpha^8 = \alpha.
\end{aligned}
$$

Table 4.2 is used to compute the error-location polynomial with the BM algorithm for the input sequence $s_0, s_1 = \alpha^5, \alpha$. The polynomial $C(D) = 1 + \alpha^3 D$ is the error-locator polynomial, and since $C(D) = \sigma(x)$ we write

$$
\sigma(x) = 1 + \alpha^3 x,
$$

whose root is found by an exhaustive search to be $\beta = \alpha^4$. The error position is the exponent of α in the representation of β^{-1}, i.e., $\beta^{-1} = \alpha^{-4} = \alpha^3$. Therefore, an error occurring at x^3 has been found, i.e., $e(x) = e_3 x^3$. In case of just a single error occurring, the error magnitude is easily found as follows:

$$
s_0 = e(\alpha) = \alpha^5 = e_3 \alpha^3,
$$

and thus $e_3 = \alpha^2$. Therefore, $e(x) = \alpha^2 x^3$ and $c(x) = r(x) - e(x) = 0$, i.e., the decoded polynomial is the all-zero polynomial. It is clear that to correct single errors it is not necessary to resort to (4.3) and (4.4).

(5) Consider the $(15, 11, 5)$ RS code over GF(16), with generator polynomial $g(x) = (x - \alpha)(x - \alpha^2)(x - \alpha^3)(x - \alpha^4)$ and decode the received polynomial $r(x) = \alpha^2 x + \alpha^3 x^9$ with the Berlekamp–Massey algorithm.

Solution: The syndrome coefficients for the received polynomial $r(x) = \alpha^2 x + \alpha^3 x^9$ are calculated with the help of Table 3.2 as

$$
\begin{aligned}
s_0 &= r(\alpha) &= \alpha^3 + \alpha^{12} &= \alpha^{10} \\
s_1 &= r(\alpha^2) &= \alpha^4 + \alpha^6 &= \alpha^{12} \\
s_2 &= r(\alpha^3) &= \alpha^5 + 1 &= \alpha^{10} \\
s_3 &= r(\alpha^4) &= \alpha^6 + \alpha^9 &= \alpha^5.
\end{aligned}
$$

Table 4.3. Evolution of the BM algorithm for the input sequence $(s_0, s_1, s_2, s_3) = (\alpha^{10}, \alpha^{12}, \alpha^{10}, \alpha^5)$, where $a(x) = 1 + \alpha^2 D + \alpha D^2$.

n	s_n	δ	$T(D)$	$C(D)$	L	$C^*(D)$	δ^*	x
0	α^{10}	α^{10}	$-$	1	0	1	1	1
1	α^{12}	α^{14}	1	$1 + \alpha^{10} D$	1	1	α^{10}	1
2	α^{10}	α^{11}	$1 + \alpha^{10} D$	$1 + \alpha^2 D$	1	1	α^{10}	2
3	α^5	α^2	$1 + \alpha^2 D$	$1 + \alpha^2 D + \alpha D^2$	2	$1 + \alpha^2 D$	α^{11}	1
			$a(x)$	$1 + \alpha^3 D + \alpha^{10} D^2$	2	$a(x)$	α^{11}	2

Table 4.3 is used to compute the error-locator polynomial with the BM algorithm for the input sequence $s_0, s_1, s_2, s_3 = \alpha^{10}, \alpha^{12}, \alpha^{10}, \alpha^5$. The polynomial $C(D) = 1 + \alpha^3 D + \alpha^{10} D^2$ is the error-locator polynomial, and since $C(D) = \sigma(x)$ we write

$$\sigma(x) = 1 + \alpha^3 x + \alpha^{10} x^2 = (1 + \alpha x)(1 + \alpha^9 x),$$

the roots of which are found by an exhaustive search to be $\beta_1 = \alpha^{14}$ and $\beta_2 = \alpha^6$. The error positions are the exponents of α in the representation of β_1^{-1} and β_2^{-1}, i.e., $\beta_1^{-1} = \alpha^{-14} = \alpha$ and $\beta_2^{-1} = \alpha^{-6} = \alpha^9$. Two errors are thus located, at positions x and x^9, respectively. The error magnitudes are found with the help of (4.3) and (4.4) as follows:

$$\begin{aligned}
Z(x) &= 1 + (s_0 + \sigma_1)x + (s_1 + \sigma_1 s_0 + \sigma_2)x^2 \\
&= 1 + (\alpha^{10} + \alpha^3)x + (\alpha^{12} + \alpha^3 \alpha^{10} + \alpha^{10})x^2 \\
&= 1 + \alpha^{12} x + \alpha^8 x^2.
\end{aligned}$$

Then

$$e_1 = \frac{Z(\beta_1)}{(1 + \beta_2^{-1}\beta_1)} = \frac{1 + \alpha^{12}\alpha^{14} + \alpha^8\alpha^{13}}{(1 + \alpha^9\alpha^{14})} = \frac{\alpha^4}{\alpha^2} = \alpha^2$$

$$e_2 = \frac{Z(\beta_2)}{(1 + \beta_1^{-1}\beta_2)} = \frac{1 + \alpha^{12}\alpha^6 + \alpha^8\alpha^{12}}{(1 + \alpha\alpha^6)} = \frac{\alpha^{12}}{\alpha^9} = \alpha^3.$$

The error polynomial is thus $\alpha^2 x + \alpha^3 x^9$. After the operation of error correction is completed, the decoded polynomial is the all-zero polynomial, i.e., $c(x) = 0$.

(6) Consider the binary $(7, 4, 3)$ BCH code having α, α^2 and α^4, as roots of the generator polynomial, where α denotes a primitive element

of GF(2^3). Decode the received word $r(x) = x^6 + x^3$ using the Euclidean frequency domain decoder.

Solution: The received word $r(x)$ in vector form is expressed as

$$\mathbf{r} = (1, 0, 0, 1, 0, 0, 0).$$

(a) The Fourier transform of $r(x)$.
    From $r(x)$ we compute R and obtain

$$R = (\alpha^2, \alpha^4, \alpha^2, \alpha, \alpha, \alpha^4, 0),$$

where

$$R_j = \sum_{i=0}^{6} r_i \alpha^{ij} = \alpha^{6j} + \alpha^{3j}.$$

(b) Syndrome calculation.
    For the syndrome we obtain $S_j = R_{j+1}, j = 0, 1$, which gives:

$$S_0 = R_1 = \alpha^4$$

$$S_1 = R_2 = \alpha,$$

and therefore,

$$S(z) = \alpha z + \alpha^4.$$

(c) Error locator calculation.
    We find $L(z)$ by means of the Euclidean algorithm (see Appendix C).

i	$g_i(z)$	$r_i(z)$	$q_i(z)$
-1	0	z^2	$-$
0	1	$\alpha z + \alpha^4$	$-$
1	$\alpha^6 z + \alpha^2$	α^6	$\alpha^6 z + \alpha^2$

therefore, $L(z) = \alpha^6 z + \alpha^2$.

(d) Error vector Fourier transform calculation.

$$\sum_{k=0}^{\nu-1} L_k E_{j-k} = 0$$

$$L_0 E_j + L_1 E_{j-1} = 0.$$

We know that $E_j = S_{j-1}$, $j = 1, 2$. It thus follows that $E_1 = \alpha^4$ and $E_2 = \alpha$. From $L(z) = \alpha^6 z + \alpha^2$ we extract $L_0 = \alpha^2$ and $L_1 = \alpha^6$, which are then applied to the recursion $L_0 E_j + L_1 E_{j-1} = 0$ to produce

$$E_j = \alpha^4 E_{j-1}.$$

We now use the recursion $E_j = \alpha^4 E_{j-1}$ to compute the remaining unknown values of $\mathbf{E}$ and obtain:

$$\mathbf{E} = (\alpha^3, \alpha^6, \alpha^2, \alpha^5, \alpha, \alpha^4, 1).$$

(e) Error correction

The inverse Fourier transform of $\mathbf{E}$ is the following:

$$\mathbf{e} = (0, 0, 1, 0, 0, 0, 0).$$

Therefore, we obtain $e(x) = x^4$, and the corrected word is given by:

$$v(x) = r(x) - e(x), \text{ i.e., } v(x) = x^6 + x^4 + x^3.$$

Chapter 5

IRREDUCIBLE POLYNOMIALS OVER FINITE FIELDS

5.1 Introduction

The theory of finite fields constitutes a very important tool in many practical situations. In particular, it is worth mentioning its use in algebraic coding and cryptography. As we have already seen, the construction of cyclic codes relies on knowledge of irreducible polynomials over finite fields. Similarly, modern cryptographic systems also rely on the structure of finite fields, e.g., the Advanced Encryption Standard (Daemen and Rijmen 2002).

Polynomials are important in the study of the algebraic structure of finite fields and irreducible polynomials are the prime elements of the polynomial ring over a finite field. Irreducible polynomials are indispensable for constructing finite fields and doing computations with elements of a finite field. The following treatment assumes the reader has familiarity with some basic properties of finite fields, some of which are now summarized.

- For every prime number p there is a finite field with p elements.

- The integers modulo p, together with ordinary addition and multiplication, are examples of finite fields.

- Let $GF(p)$ denote a finite field with p elements. Given an irreducible polynomial of degree m over $GF(p)$, a field with p^m elements can always be constructed.

- There can be no finite field with a number of elements which is not a power of a prime number.

A very important result in finite field theory, which is proven later in Theorems 5.10 and 5.11, is explained as follows. For all values of a prime number p and a positive integer m there exists at least one irreducible polynomial of degree m over $\mathrm{GF}(p)$. Also, the fields with p^m elements, constructed with such irreducible polynomials, are the only kind of finite fields that exist, i.e., they are unique up to an isomorphism. Those results are established by analyzing how certain polynomials factor over certain finite fields.

5.2   Order of a polynomial

LEMMA 5.1 *Let $f(x) \in \mathrm{GF}(q)[x]$ be a polynomial of degree $m \geq 1$ with $f(0) \neq 0$. Then there exists a positive integer $e \leq q^m - 1$ such that $f(x)|(x^e - 1)$.*

Proof: The residue class ring $\mathrm{GF}(q)[x]/(f(x))$ contains $q^m - 1$ nonzero residue classes, which are obtained from the nonzero remainders modulo $f(x)$ of polynomials over $\mathrm{GF}(q)[x]$. The $q^m - 1$ residue classes $x^j + (f(x))$, $j = 1, 2, \ldots, q^m - 1$, are all nonzero and must be included among the nonzero residue classes given earlier, so there exist integers r and s with $0 \leq r < s \leq q^m - 1$ such that $x^s \equiv x^r$ modulo $f(x)$. Since x and $f(x)$ are relatively prime, i.e., their greatest common divisor is 1, denoted by $\gcd(x, f(x)) = 1$, or simply as $(x, f(x)) = 1$, it follows that $x^{s-r} \equiv 1$ modulo $f(x)$, i.e., $f(x)$ divides $x^{s-r} - 1$, and $0 < s - r \leq q^m - 1$.

$\square$

DEFINITION 5.2 *Let $f(x) \in \mathrm{GF}(q)[x]$ be a nonzero polynomial. If $f(0) \neq 0$, then the least positive integer e for which $f(x)$ divides $x^e - 1$ is called the order of $f(x)$.*

The order of $f(x)$ is denoted by $\mathrm{ord}(f) = \mathrm{ord}(f(x))$. If $f(0) = 0$, then $f(x) = x^h g(x)$, $h \in \mathbb{N}$ and $g(x) \in \mathrm{GF}(q)[x]$, with $g(0) \neq 0$, are uniquely determined. In this case, the $\mathrm{ord}(f) = \mathrm{ord}(g)$ by definition. The order of $f(x)$ is also known as the period of $f(x)$ or the exponent to which $f(x)$ belongs.

5.3   Factoring $x^{q^n} - x$

Consider a finite field $\mathrm{GF}(q)$, $q = p^m$ and the polynomial $x^{q^n} - x$. Such a polynomial has a unique factorization as a product of irreducible monic polynomials over $\mathrm{GF}(q)$. This result can be expressed in a slightly different form as given in Theorem 5.4, but first the following lemma is needed.

LEMMA 5.3 *Let α be a primitive element of $\mathrm{GF}(q^d)$ such that $\alpha^{q^s} = \alpha$. Any element $\beta \in \mathrm{GF}(q^d)$ is a root of $x^{q^s} - x$, as long as $s \geq d$.*

Proof: Let $\beta \in \mathrm{GF}(q^d)$ and let $a_i \in \mathrm{GF}(q)$, then $\beta = \sum_{i=0}^{d-1} a_i \alpha^i \to \beta^{q^s} = \sum_{i=0}^{d-1} a_i^{q^s} \alpha^{iq^s} = \sum_{i=0}^{d-1} a_i \alpha^i = \beta$, since $\alpha^{q^s} = \alpha$ and $a_i^{q^s} = a_i$. Thus, any element $\beta \in \mathrm{GF}(q^d)$ is a root of $x^{q^s} - x$, as long as $s \geq d$. □

THEOREM 5.4 *The polynomial $x^{q^n} - x$ factors over $\mathrm{GF}(q)$ as the product*

$$\prod_{d:d|n} V_d(x),$$

where $V_d(x)$ is the product of all monic irreducible polynomials of degree d in $\mathrm{GF}(q)[x]$, and where d is a divisor of n, i.e.,

$$x^{q^n} - x = \prod_{d:d|n} V_d(x). \tag{5.1}$$

Proof: First we prove that $x^{q^n} - x$ has no repeated factor, and then we prove that the degree of every irreducible divisor of $x^{q^n} - x$ is a divisor of n. To show that $x^{q^n} - x$ has no repeated factor, it is necessary to calculate the greatest common divisor (gcd) between $x^{q^n} - x$ and its formal derivative (which is equal to -1) to check whether they are relatively prime. It follows that $\gcd(x^{q^n} - x, -1) = 1$ and consequently $x^{q^n} - x$ has no repeated factor. Now let $f(x)$ be a monic irreducible polynomial of degree d, $d|n$, over $\mathrm{GF}(q)$. Let α be a root of $f(x)$, i.e., let $f(\alpha) = 0$. An extension field $\mathrm{GF}(q^d)$, constructed with $f(x)$, has q^d elements and thus

$$\alpha^{q^d} = \alpha, \tag{5.2}$$

which is equivalent to

$$f(x)|(x^{q^d} - x).$$

However, $(x^{q^d} - x)|(x^{q^n} - x)$ if and only if $d|n$, and it follows that $f(x)|(x^{q^n} - x)$ because, by hypothesis, $d|n$. Thus, every irreducible polynomial the degree of which divides n is a factor of $(x^{q^n} - x)$. The converse is proven next by assuming that $f(x)|(x^{q^n} - x)$. If the degree of $f(x)$ is d, (5.2) still holds and so $f(x)| \gcd\left[(x^{q^d} - x), (x^{q^n} - x)\right] = x^{q^s} - x$, $s = \gcd(d, n)$, and since $f(\alpha) = 0$ then necessarily $\alpha^{q^s} = \alpha$ must hold. In fact, by Lemma 5.3 any element $\beta \in \mathrm{GF}(q^d)$ is a root of $x^{q^s} - x$ which can have at most q^s solutions, and so $s \geq d$. Since $s = \gcd(n, d)$, it follows that $s = d$. This means that d is a divisor of n. □

EXAMPLE 5.5 *Let $q = 2$ and $n = 3$. By Theorem 5.4, $x^{2^3} + x$ is the product of all GF(2)-irreducible polynomials of degree d, $d|3$, i.e., of degrees 1 and 3. From a table of irreducible polynomials, e.g., (Peterson and Weldon Jr. 1972, p.472) it follows that*

$$x^8 + x = x(x + 1)(x^3 + x + 1)(x^3 + x^2 + 1)$$

and thus

$$
\begin{aligned}
V_1(x) &= x^2 + x \\
V_3(x) &= x^6 + x^5 + x^4 + x^3 + x^2 + x + 1.
\end{aligned}
$$

Comparing the degrees on both sides of (5.1) the following corollary is obtained.

COROLLARY 5.6 *$q^n = \sum_{d:d|n} d I_d$, where I_d is the number of distinct monic irreducible polynomials of degree d.*

5.4 Counting monic irreducible q-ary polynomials

The powerful tool of generating functions is now used to obtain an analytic expression for the unique factorization theorem for polynomials over a finite field. Let the generating function $M(z)$ be defined as

$$M(z) = \sum_{n \geq 0} T_n z^n, \tag{5.3}$$

where T_n denotes the number of monic polynomials of degree n over GF(q). The proof consists of calculating (5.3) in two distinct ways and equating the results.

WAY 1: In a general monic polynomial $f(x) = x^n + \sum_{i=0}^{n-1} a_i x^i$, of degree n over GF(q), there are q distinct choices for each one of the coefficients a_i of x^i, $0 \leq i \leq n-1$. Therefore, there are q^n such polynomials, i.e., $T_n = q^n$. From (5.3) it follows that

$$M(z) = \sum_{n \geq 0} q^n z^n = \frac{1}{1 - qz}. \tag{5.4}$$

WAY 2: Over a field every monic polynomial of degree n has a unique factorization into a product of monic irreducible polynomials. The following generating function (Berlekamp 1968, p.76), called *enumerator by degree* is now used to represent the number of powers of a particular monic irreducible polynomial $p(x)$ of degree d, i.e.,

$$1 + z^d + z^{2d} + \cdots = \frac{1}{1 - z^d}.$$

The power series $1+z^d+z^{2d}+\cdots$ is interpreted as follows. The coefficient of a term z^i, in the series, is equal to 1 if i is a multiple of d, i.e., if it corresponds to a polynomial of degree i which is a power of $p(x)$, otherwise the coefficient of z^i is made equal to 0. Since for a given d there are I_d irreducible monic polynomials of degree d, the expression

$$\prod_{d\geq 1}\left(\frac{1}{1-z^d}\right)^{I_d},\tag{5.5}$$

can be written to represent the number of all monic polynomials which are products of powers of irreducible monic polynomials. This covers all possible monic polynomials and thus (5.4) and (5.5) can be equated, i.e.,

$$\frac{1}{1-qz}=\prod_{d\geq 1}\left(\frac{1}{1-z^d}\right)^{I_d}.\tag{5.6}$$

Expression (5.6) represents the unique factorization theorem analytically.

Alternative proof of Corollary 5.6

Proof: Taking logarithms on both sides in (5.6), then taking the derivative of both sides, and multiplying both sides by z, it follows that

$$\frac{qz}{1-qz}=\sum_{d\geq 1}dI_d\left(\frac{z^d}{1-z^d}\right).\tag{5.7}$$

However, the right-hand side of (5.7) can be written as

$$\sum_{d\geq 1}dI_d\left(\frac{z^d}{1-z^d}\right)=\sum_{d\geq 1}dI_d\sum_{m\geq 1}z^{md}=\sum_{n\geq 1}z^n\sum_{d|n}dI_d$$

and since $qz/(1-qz)=\sum_{n\geq 1}q^nz^n$, Corollary 5.6 results by equating the coefficients of z^n on both sides of (5.7). $\square$

5.5 The Moebius inversion technique

By looking at Theorem 5.4 and Corollary 5.6 the following expressions are worth noticing:

$$x^{q^n}-x=\prod_{d:d|n}V_d(x)\tag{5.8}$$

$$q^n=\sum_{d:d|n}dI_d.\tag{5.9}$$

How can the values of $V_d(x)$ and I_d, in (5.8) and (5.9), respectively, be extracted from their associated expressions? At the present stage, nothing much better than a brute force computation of these values can be done. The neat answer to this question is provided by the Moebius inversion formula, which is investigated next.

5.5.1 The additive Moebius inversion formula

Consider G, an abelian group, with group operation $+$. Suppose that $a(1), a(2), \ldots$ and $b(1), b(2), \ldots$ are two sequences of elements in G which are related by

$$a(n) = \sum_{d:d|n} b(d), \ n \geq 1. \tag{5.10}$$

It is desired to find a formula to express the b's as a function of the a's. This is called the problem of inverting (5.10). First notice that (5.10) uniquely determines $b(d)$ as a function of $a(n)$, since $b(1) = a(1)$ for $n = 1$, and for $n \geq 1$ it follows that

$$b(n) = a(n) - \sum_{d:d|n} b(d), \ d \neq n.$$

The sequence $\mu(n)$, known as the Moebius function, is now defined to be the inverse of the sequence $(1, 0, 0, \ldots)$ and, as shown shortly, this will prove useful in solving the problem of inverting (5.10), i.e.,

$$\sum_{d:d|n} \mu(d) = \begin{cases} 1, & \text{if } n = 1 \\ 0, & \text{if } n > 1. \end{cases} \tag{5.11}$$

Actually, the desired formula is the following:

$$b(n) = \sum_{d:d|n} a(d)\mu(n/d). \tag{5.12}$$

Doing it this way is much easier than starting from scratch (Berlekamp 1968, p.81). Taking this value of $b(n)$ into (5.10) it follows that

$$
\begin{aligned}
a(n) &= \sum_{d:d|n} b(d) = \sum_{d:d|n} \sum_{e:e|d} a(e)\mu(d/e) \\
&= \sum_{e:e|n} a(e) \sum_{d:e|d|n} \mu(d/e) = \sum_{e:e|n} a(e) \sum_{f|(n/e)} \mu(f) = a(n)
\end{aligned}
$$

by the definition of $\mu(f)$, where $f = d/e$.

5.5.1.1   Computation of $\mu(n)$

The computation of $\mu(n)$ by using its definition in (5.11) is cumbersome. A simpler way to compute $\mu(n)$ is presented next. Let $n = p_1^{e_1} p_2^{e_2} \ldots p_m^{e_m}$ be the factorization of n into distinct prime powers. Then the following definition of $\mu(n)$:

$$\mu(n) = \begin{cases} 1, & \text{if } n = 1 \\ (-1)^m, & \text{if } n \text{ is the product of } m \text{ distinct primes} \\ 0, & \text{if any } e_i \geq 2 \end{cases} \quad (5.13)$$

is next shown to be the same as the definition in (5.11).

Let $x_1, x_2, \ldots, x_m$ be indeterminates and let $p_1^{e_1} p_2^{e_2} \ldots p_m^{e_m}$ be the factorization of n into distinct prime powers. For each d, $d|n$, d is factored as a product containing powers of some of the p_i's. $x(d)$ is defined as the product of x_i's, where the i's are the same as the subscripts in the p_i's which divide d, i.e.,

$$x(d) = \prod_{p_i | d} x_i$$

with $x(1) = 1$. Example 5.7 is meant to clarify the definition of $x(d)$.

EXAMPLE 5.7 *Let* $n = p_1^3 p_2 p_3^2$ *and consider the following: values of d,* $d|n$.

(a) $d = p_1 p_2 p_3$,

(b) $d = p_1^2 p_2$,

(c) $d = p_1^3 p_3^2$,

(d) $d = p_1 p_3^2$.

The corresponding values of $x(d)$ are, respectively,

(a) $x(p_1 p_2 p_3) = x_1 x_2 x_3$

(b) $x(p_1^2 p_2) = x_1 x_2$

(c) $x(p_1^3 p_3^2) = x_1 x_3$

(d) $x(p_1 p_3^2) = x_1 x_3$.

The product $\prod_{i=1}^{m}(1 - x_i)$ is now expanded, and in view of (5.13) it is observed that

$$\prod_{i=1}^{m}(1 - x_i) = \sum_{d:d|n} \mu(d)x(d). \quad (5.14)$$

Making $x_i = 1$, $1 \le i \le m$, in (5.14) it follows that

$$\sum_{d:d|n} \mu(d) = \begin{cases} 1, & \text{if } n = 1 \\ 0, & \text{if } n > 1, \end{cases}$$

and since this result is in agreement with the definition in (5.11) the formula given in (5.13) is correct. This is now stated as a theorem.

THEOREM 5.8 *If $a(n)$ and $b(n)$ are two sequences of elements in a commutative group G, which satisfy the relation given in (5.10) then expression (5.10) can be inverted to give*

$$b(n) = \sum_{d:d|n} a(d)\mu(n/d), \tag{5.15}$$

where the function $\mu(n)$ is as given in (5.13).

By making $c = n/d$ in (5.15) an alternative expression is obtained for the additive Moebius inversion formula, namely

$$b(n) = \sum_{c:c|n} a(n/c)\mu(c). \tag{5.16}$$

Expression (5.15) is called the additive Moebius inversion formula, and the function $\mu(n)$ is called the Moebius function.

5.5.2   The multiplicative Moebius inversion formula

If the underlying Abelian group operation is multiplication then the following version of the Moebius inversion formula is appropriate.

THEOREM 5.9 *If $a(n) = \prod_{d:d|n} b(d)$ then*

$$b(n) = \prod_{d:d|n} a(d)^{\mu(n/d)} \tag{5.17}$$

or

$$b(n) = \prod_{d:d|n} a(n/d)^{\mu(d)}.$$

Proof: Taking the value of $b(n)$ into the expression for $a(n)$ it follows that

$$a(n) = \prod_{d:d|n} b(d) = \prod_{d:d|n} \prod_{e:e|d} a(e)^{\mu(d/e)} = \prod_{e:e|n} \prod_{d:e|d|n} a(e)^{\mu(d/e)}$$

$$= \prod_{e:e|n} a(e)^{\sum_{f|(n/e)} \mu(f)} = a(n),$$

where $f = d/e$. $\qquad\qquad\square$

In order to motivate the present study some interesting applications of (5.15) are presented next, related to some of the questions raised earlier.

5.5.2.1   The Euler totient function

The number of integers in the set $\{0, 1, 2, \ldots, t-1\}$ which are relatively prime to t is represented by the symbol $\phi(t)$. The symbol $\phi(.)$ is called the Euler ϕ function or the Euler totient function. In number theory, it is shown that if n is any positive integer, then $\sum_{d|n} \phi(d) = n$. Next, Theorem 5.8 is applied to this formula, by identifying $b(d) = \phi(d)$ and $a(n) = n$. Thus, from (5.16) it follows that

$$\phi(n) = \sum_{d:d|n} (n/d)\mu(d) = n \sum_{d:d|n} \mu(d)/d. \tag{5.18}$$

The formula (5.18) still does not look like a simple way to compute $\phi(n)$. Suppose that n factors into distinct prime powers as

$$n = p_1^{e_1} p_2^{e_2} \cdots p_m^{e_m}.$$

Now applying (5.14) with $x_i = 1/p_i$, and noticing that $x(d) = 1/d$, it follows that

$$\sum_{d:d|n} \frac{\mu(d)}{d} = \prod_{i=1}^{m} (1 - 1/p_i). \tag{5.19}$$

By combining (5.18) and (5.19) finally it follows that

$$\phi(n) = n \prod_{i=1}^{m} (1 - 1/p_i) = \prod_{i=1}^{m} p_i^{e_i-1}(p_i - 1). \tag{5.20}$$

5.5.3   The number of irreducible polynomials of degree n over GF(q)

An expression is derived next for calculating the number of irreducible polynomials of a given degree over a given finite field. From Corollary 5.6, it is known that $q^n = \sum_{d:d|n} d I_d$. Let $a(n) = q^n$ and let $b(n) = n I_n$ in (5.10). The underlying group is formed by the integers under ordinary addition. Now, applying (5.16) with these values it follows that

$$I_n = \frac{1}{n} \sum_{d:d|n} \mu(d) q^{n/d}. \tag{5.21}$$

The formula in (5.21) gives the number of irreducible polynomials of degree n over a finite field with q elements. It is a very important result in itself and in the conclusions that can be drawn about the existence of irreducible polynomials. For example, since the dominant term in (5.21) is obtained when $d = 1$, I_n can be approximated as $I_n \approx q^n/n$, for large n and q fixed. This can be interpreted to mean that, for large n and q fixed, the probability of randomly choosing an irreducible polynomial of degree n is about $1/n$, because $I_n/q^n \approx 1/n$. In (McEliece 1987, p.66) the author remarks that, although interesting, that is by no means a proof of the existence of an irreducible polynomial of any degree.

THEOREM 5.10 *For a given finite field with q elements, $\mathrm{GF}(q)$, for all $n \geq 1$ there exists at least one irreducible polynomial of degree n over $\mathrm{GF}(q)$.*

Proof: Notice that $I_1 = q$. Also, since $I_t \geq 0$ for all t, it follows that

$$q^t = \sum_{d:d|t} dI_d = tI_t + \sum_{d:d|t,d\neq t} dI_d = tI_t + \sum_{d:d|t,d\neq t,d\neq 1} dI_d + I_1,$$

where $I_1 = q$, and it follows that

$$q^t \geq tI_t + I_1 \text{ or } q^t \geq tI_t + q.$$

Thus, $I_t \leq (q^t - q)/t$, with equality if and only if t is prime. In particular, $I_2 = (q^2 - q)/2 > 0$ and $I_3 = (q^3 - q)/3 > 0$. Next, a lower bound on I_t is derived. Since $q^t \geq tI_t$, it follows that

$$\sum_{i=0}^{t/2} q^i > \sum_{i=0}^{t/2} iI_i \geq \sum_{d:d|t,d\neq t} dI_d \tag{5.22}$$

and thus,

$$q^t = \sum_{d:d|t} dI_d = tI_t + \sum_{d|t,d\neq t} dI_d < tI_t + \sum_{i=0}^{t/2} q^i < tI_t + q^{t/2+1},$$

where the first inequality results from (5.22) and the second inequality results because $\sum_{i=0}^{t/2} q^i$ is part of a *super increasing* sequence (Denning 1982), i.e., each term in the sequence is greater than the sum of the preceding terms. In this manner, we obtain the inequality

$$I_t > (q^t - q^{t/2+1})/t = (1 - q^{-t/2+1})q^t/t.$$

This last expression shows that $I_t > 0$ for all $t \geq 2$. It has already been seen that $I_1 > 0$ and $I_2 > 0$, and thus it is concluded that there exist irreducible monic polynomials of every degree over every finite field.   □

As a consequence of Theorem 5.10 the following important result is obtained.

THEOREM 5.11 *If p is any prime number and m is any positive integer, then there exists a finite field having p^m elements.*

5.6 Chapter citations

Chapters 3, 4, and 6 of (Berlekamp 1968), Chapters 2, 3, and 4 of (Lidl and Niederreiter 2006), Chapters 6 and 7 of (McEliece 1987) and Chapters 3 and 4 of (MacWilliams and Sloane 1977).

5.7 Problems with solutions

(1) Prove that $(x^{q^d} - x)|(x^{q^n} - x)$ if and only if $d|n$.

Solution: Simplifying the given polynomial ratio we obtain

$$F(x) = \frac{x^{q^n} - x}{x^{q^d} - x} = \frac{x^N - 1}{x^D - 1},$$

where $N = q^n - 1$ and $D = q^d - 1$. Suppose, with no loss of essential generality, that $N = aD + R$, $R < D$. Then we can express $F(x)$ as

$$F(x) = x^R \left(\frac{x^{aD} - 1}{x^D - 1} \right) + \frac{x^R - 1}{x^D - 1}.$$

We know that $x^{aD} - 1$ is always divisible by $x^D - 1$ (Prove it!) and since $R < D$, $x^R - 1$ is divisible by $x^D - 1$ if and only if $R = 0$, i.e., if and only if $D|N$. By a similar argument it follows that $d|n$, i.e., $(q^d - 1)|(q^n - 1)$ if and only if $d|n$.

(2) Prove that $\gcd\{(x^{q^d} - x), (x^{q^n} - x)\} = x^{q^s} - x$, where $s = \gcd(n, d)$.

Solution: We know by Theorem 5.4 that

$$x^{q^n} - x = \prod_{a: a|n} V_a(x)$$

and that

$$x^{q^d} - x = \prod_{b: b|d} V_b(x).$$

Thus, $\gcd(x^{q^n} - x, x^{q^d} - x)$ is the product of all monic irreducible polynomials whose degrees divide both n and d, i.e., whose degrees divide $\gcd(n, d) = s$. Again from Theorem 5.4 we know that the product of all polynomials whose degrees divide s is equal to $x^{q^s} - 1$. Thus,

$$\gcd(x^{q^n} - x, x^{q^d} - x) = x^{q^s} - x,$$

where $s = \gcd(n, d)$.

(3) Let $n = 6$ in (5.14). Expand both sides and verify that they coincide.

Solution: We consider (5.14) as follows:

$$\prod_{i=1}^{m}(1 - x_i) = \sum_{d|n} \mu(d)x(d),$$

where $x(1) = 1$, $n = p_1^{e_1}p_2^{e_2}\ldots p_m^{e_m}$ and, e.g., $x(p_i^{e_i}p_j^{e_j}\ldots p_l^{e_l}) = x_i x_j \ldots x_l$.

(a) For $n = 6 = 2.3$, expanding the left-hand side of (5.14) we obtain

$$(1 - x_1)(1 - x_2) = 1 - x_1 - x_2 + x_1 x_2.$$

(b) For $n = 6 = 2.3$, expanding the right-hand side of (5.14) we obtain

$$\mu(1)x(1) + \mu(2)x(2) + \mu(3)x(3) + \mu(6)x(6)$$
$$= 1.1 - 1.x_1 - 1.x_2 + 1.x_1 x_2,$$

and, as expected, the results obtained in (a) and (b) coincide.

(4) Verify (5.19) for the particular value $n = 12$.

Solution: We consider (5.19) as follows:

$$\sum_{d|n} \frac{\mu(d)}{d} = \prod_{i=1}^{m}\left(1 - \frac{1}{p_i}\right),$$

where $n = p_1^{e_1}p_2^{e_2}\ldots p_m^{e_m}$.

a) For $n = 12$, expanding the left-hand side of (5.19) we obtain

$$\sum_{d|12} \frac{\mu(d)}{d} = \mu(1) + \frac{\mu(2)}{2} + \frac{\mu(3)}{3} + \frac{\mu(4)}{4} + \frac{\mu(6)}{6} + \frac{\mu(12)}{12}$$

$$= 1 - \frac{1}{2} - \frac{1}{3} + \frac{0}{4} + \frac{1}{6} + \frac{0}{12}$$

$$= 1 - \frac{1}{2} - \frac{1}{3} + \frac{1}{6}.$$

b) For $n = 12$, expanding the right-hand side of (5.19) we obtain

$$\prod_{i=1}^{2}\left(1 - \frac{1}{p_i}\right) = \left(1 - \frac{1}{2}\right)\left(1 - \frac{1}{3}\right)$$

$$= 1 - \frac{1}{2} - \frac{1}{3} + \frac{1}{6},$$

and, as expected, the results obtained in (a) and (b) coincide.

(5) Obtain the enumerators by degree for each of the following subsets of monic polynomials over $GF(q)$:

(a) Perfect squares.

(b) Perfect cubes.

(c) Products of all distinct irreducible factors.

Solution:

(a) *Perfect squares*: By squaring a monic polynomial of degree k we obtain a perfect square of degree $2k$. Therefore, the number of perfect squares of degree $2k$ coincides with the number of monic polynomials of degree k, i.e., q^k. The total number of distinct perfect squares is written in terms of the enumerator by degree (see Section 5.4) as

$$\sum_{k=0}^{\infty} q^k z^{2k} = \sum_{k=0}^{\infty} (qz^2)^k = \frac{1}{1 - qz^2}.$$

(b) *Perfect cubes*: Following a reasoning analogous to the one in the listed in (a) we obtain

$$\sum_{k=0}^{\infty} q^k z^{3k} = \sum_{k=0}^{\infty} (qz^3)^k = \frac{1}{1 - qz^3}.$$

(c) Products of all distinct irreducible factors: Assuming $q = 2$, for $k = 1$ we can easily check that x and $x+1$ are the only degree one distinct irreducible polynomials. For $k = 2$ we obtain $x(x + 1)$ and $x^2 + x + 1$. For $k = 3$ we obtain $x(x^2 + x + 1), (x + 1)(x^2 + x + 1), x^3 + x^2 + 1$, and $x^3 + x + 1$. For $k \geq 2$ we can check that 2^{k-1} distinct irreducible factors result. In general, over $GF(q)$ we have the following expression:

$$\sum_{k=0}^{\infty} q^k z^k - \sum_{k=2}^{\infty} q^{k-1} z^k, \tag{5.23}$$

which denotes the total number of monic polynomials minus the number of polynomials which contain at least one factor with multiplicity 2 or greater. Alternately, (5.23) can be written as

$$\sum_{k=0}^{\infty} q^k z^k - qz^2 \sum_{j=0}^{\infty} q^j z^j = \frac{1}{1-qz} - \frac{qz^2}{1-qz} = \frac{1-qz^2}{1-qz}.$$

(6) Let $g(x)$ be an irreducible polynomial with nonzero derivative $g'(x)$. Show that $g(x)$ is a repeated factor of the polynomial $f(x)$ if and only if $g(x)$ divides $\gcd(f(x), f'(x))$ where $f'(x)$ is the formal derivative of $f(x)$.

REMARK (Berlekamp 1968, p.75), (Castagnoli, Massey, and Schoeller 1991): In a finite field or in a field of characteristic zero, every irreducible polynomial has a nonzero derivative, however, in certain infinite fields with prime characteristic there exist irreducible polynomials the derivative of which is zero.

Solution: Let $g(x)$ be an irreducible factor of $f(x)$, i.e., let $f(x) = g(x)h(x)$. The formal derivative of $f(x)$ produces

$$f'(x) = g'(x)h(x) + g(x)h'(x).$$

If $g(x)$ divides $f'(x)$ then it must divide $g'(x)h(x)$. However, $\deg g'(x) < \deg g(x)$ and $g(x)$ does not divide $g'(x)$, thus $g(x)$ divides $h(x)$. Therefore $g(x)$ is a repeated factor of $f(x)$ if and only if it is a factor of both $f(x)$ and $f'(x)$, or, in other words, if it divides $\gcd(f(x), f'(x))$.

(7) Show that $\sum_{d:d^2|n} \mu(d) = |\mu(n)|$.

Solution:

(a) Let n be a prime. Then $\mu(n) = -1$ and $\sum_{d:d^2|n} \mu(d) = \mu(1) = 1$. So, $\sum_{d:d^2|n} \mu(d) = |\mu(n)|$ in this case.

(b) Let n be a product of non-repeated primes. Then $\mu(n) = (-1)^m$ if $n = p_1 p_2 \ldots p_m$ and $\sum_{d:d^2|n} \mu(d) = \mu(1) = 1$, thus, $\sum_{d:d^2|n} \mu(d) = |\mu(n)|$.

(c) The general case.
Let $p_1^{e_1} p_2^{e_2} \ldots p_m^{e_m}$ be the prime factor decomposition of n. Now, we have

$$\sum_{d:d^2|n} \mu(d) = \mu(1) + \mu(p_{l_1}) + \mu(p_{l_2}) + \cdots + \mu(p_{l_s}) +$$
$$\mu(p_{l_1} p_{l_2}) + \mu(p_{l_1} p_{l_3}) + \cdots + \mu(p_{l_{s-1}} p_{l_s}) +$$
$$\cdots + \mu(p_{l_1} \ldots p_{l_{s-1}} p_{l_s}),$$

where each p_{l_j}, $1 \le j \le s$, contains at least one factor $p_i^{e_i}$ with $e_i \ge 2$, for $1 \le i \le m$. It thus follows that $\sum_{d:d^2|n} \mu(d) = \mu(1) = 1$ since all the remaining terms in the sum will be zero. The result is thus true in general.

(8) Evaluate the sum $\sum_{d|n} |\mu(d)|$.

Solution: Let m be the number of distinct prime factors of n. Consider the relation

$$\prod_{i=1}^{m}(1 - x_i) = \sum_{d:d|n} \mu(d)x(d) = \mu(1) + \mu(p_1)x_1 + \mu(p_2)x_2 +$$
$$+ \cdots + \mu(p_1 p_2 \ldots p_m)x_1 x_2 \ldots x_m \tag{5.24}$$

obtained from (5.14), and make $x_i = -1$ for all i. It now follows that

$$2^m = \sum_{d:d|n} |\mu(d)|$$

because any nonzero term in (5.24) becomes positive, e.g.,

$$\mu(p_1 p_2 \ldots p_s)x_1 x_2 \ldots x_s = (-1)^s(-1)^s = (-1)^{2s} = 1,$$

for $x_i = -1, 1 \le i \le s$, and there are 2^m such terms.

(9) Show that the generating function $A(x) = \sum_{i=0} A_i x^i$ has a multiplicative inverse, denoted by $B(x)$ such that $A(x)B(x) = 1$, if and only if $A_0 \ne 0$. *Hint:* Equate coefficients of the same degree on both sides of $A(x)B(x) = 1$.

Solution:

$$A(x) = \sum_{i=0}^{\infty} A_i x^i.$$

If $A_0 \ne 0$ we can write

$$\frac{A(x)}{A_0} = 1 + \sum_{i=1}^{\infty} \left(\frac{A_i}{A_0}\right) x^i$$

and

$$\frac{A_0}{A(x)} = \frac{1}{1 + \sum_{i=1}^{\infty} \left(\frac{A_i}{A_0}\right) x^i} = \sum_{j=0}^{\infty} \left[-\sum_{i=1}^{\infty} \left(\frac{A_i}{A_0}\right) x^i\right]^j.$$

Thus

$$B(x) = \frac{1}{A(x)} = \frac{1}{A_0} \sum_{j=0}^{\infty} \left[-\sum_{i=1}^{\infty} \left(\frac{A_i}{A_0} \right) x^i \right]^j$$

is defined if and only if $A_0 \neq 0$.

Chapter 6

FINITE FIELD FACTORIZATION OF POLYNOMIALS

6.1   Introduction

The development and design of many practical applications involving digital sequences require the factorization of polynomials. In particular, a cyclic code or a similar mathematical structure requires the factorization of $x^n - 1$ as a product of irreducible polynomials with roots in a finite field, where n denotes a positive integer. The global positioning system (Parkinson and Spilker Jr. 1996, pp.114–118) and digital video broadcast (Alencar 2009, pp.89–93) are just two examples where the factorization of polynomials in finite fields is employed.

Let again the factorization derived in Theorem 5.4 be considered, i.e.,

$$x^{q^n} - x = \prod_{d:d|n} V_d(x). \tag{6.1}$$

In (6.1) $V_d(x)$ denotes the product of all monic irreducible polynomials of degree d, over $\mathrm{GF}(q)$, such that $d|n$. We want to extract $V_d(x)$ in (6.1) and to do that the multiplicative Moebius inversion technique will be employed over the Abelian group formed by the set of rational functions with ordinary multiplication as the group operation. By *the set of rational functions over* $\mathrm{GF}(q)$ is meant the set

$$\{p(x)/q(x),\ p(x) \in \mathrm{GF}(q)[x],\ q(x) \in \mathrm{GF}(q)[x],\ p(x) \neq 0,\ q(x) \neq 0\}.$$

The multiplicative Moebius inversion formula (5.17) is represented as

$$b(n) = \prod_{d:d|n} [a(d)]^{\mu(n/d)}, \tag{6.2}$$

equation where

$$[a(d)]^0 \;=\; a^0(d) \;=\; 1 \text{ is the group identity,}$$
$$[a(d)]^{+1} \;=\; a^{+1}(d) \;=\; a(d)$$

$$[a(d)]^{-1} \;=\; a^{-1}(d) \;=\; \text{the multiplicative inverse of } a(d).$$

Expressions (6.1) and (6.2) are combined by making $b(n) \Leftarrow V_n(x)$ and $a(d) \Leftarrow x^{q^d} - x$, and the result is the following:

$$V_n(x) = \prod_{d:d|n} (x^{q^d} - x)^{\mu(n/d)}. \tag{6.3}$$

EXAMPLE 6.1 *Let $q = 2$ and let $n = 10$ in (6.3). It follows that*

$$
\begin{aligned}
V_{10}(x) &= \prod_{d:d|10} (x^{2^d} - x)^{\mu(10/d)} \\
&= (x^2 - x)^{\mu(10)}(x^4 - x)^{\mu(5)}(x^{32} - x)^{\mu(2)}(x^{1024} - x)^{\mu(1)} \\
&= \frac{(x^2 - x)(x^{1024} - x)}{(x^4 - x)(x^{32} - x)} = \frac{(x-1)(x^{1023} - 1)}{(x^3 - 1)(x^{31} - 1)},
\end{aligned}
$$

i.e., $V_{10}(x)$ is a polynomial of degree 990 which is not yet known explicitly. Since $990/10 = 99$, it follows from Theorem 5.4 that $V_{10}(x)$ is the product of 99 monic irreducible polynomials of degree 10 over GF(2). For the moment that is as far as one can go, concerning the factorization of $V_{10}(x)$. To proceed with this factorization more theory needs to be introduced.

6.2   Cyclotomic polynomials

Consider GF(q) a finite field of characteristic p. Let n be a positive integer, $\gcd(n, p) = 1$, and let ζ be a primitive nth root of unity, i.e., $\zeta^u \neq 1$, $1 \leq u < n$ and $\zeta^n = 1$. Over GF(q), the following factorization results

$$x^n - 1 = \prod_{u=0}^{n-1} (x - \zeta^u). \tag{6.4}$$

It is remarked that the order of ζ^u depends on $\gcd(n, u)$ and that, for every divisor d of n, there are $\phi(d)$ powers of ζ that have order d.

DEFINITION 6.2 *The dth cyclotomic polynomial $\Phi_d(x)$ is the monic polynomial of degree $\phi(d)$ which has as its roots those powers of ζ which have order d, $d|n$, where ζ is a primitive nth root of unity over GF(q), i.e.,*

$$\Phi_d(x) = \prod_{u=0}^{n-1} (x - \zeta^u), \quad \gcd(n, u) = n/d. \tag{6.5}$$

By combining (6.4) and (6.5) it follows that

$$x^n - 1 = \prod_{d:d|n} \Phi_d(x).$$ (6.6)

By applying the Moebius inversion formula to (6.6), with $b(n) \Leftarrow \Phi_n(x)$ and $a(d) \Leftarrow x^d - 1$, it follows that

$$\Phi_n(x) = \prod_{d:d|n} (x^d - 1)^{\mu(n/d)}.$$ (6.7)

By expanding the right-hand side of (6.7) a quotient of two monic polynomials is obtained, where the numerator is the product of those $(x^d - 1)$ for which $\mu(n/d) = 1$ and where the denominator is the product of those $(x^d - 1)$ for which $\mu(n/d) = -1$. Notice that $\Phi_n(x)$ in (6.7) has integer coefficients, despite the appearance of GF(q) elements in Definition 6.2, because (6.7) is the quotient of two polynomials where the numerator has integer coefficients and the denominator is *monic* with integer coefficients.

EXAMPLE 6.3 *Compute* $\Phi_{12}(x)$.

Solution: *By (6.7) it follows that*

$$\Phi_{12}(x) = \frac{(x^{12} - 1)(x^2 - 1)}{(x^6 - 1)(x^4 - 1)}.$$ (6.8)

By performing the operations indicated on the right-hand side of (6.8) a degree 4 polynomial results. However, it was noticed (Berlekamp 1968, p.95) that, since $\deg \Phi_d(x) = \phi(d)$, *the operations needed to obtain* $\Phi_n(x)$ *may be simplified by reducing polynomials modulo* x^t, $t = \phi(n) + 1$. *Thus it follows that*

$$
\begin{aligned}
\Phi_{12}(x) &= \frac{(-1)(x^2 - 1)}{(-1)(x^4 - 1)} \quad \text{mod } x^5 \\
&= \frac{(x^2 - 1)(x^4 + 1)}{(x^4 - 1)(x^4 + 1)} \equiv x^4 - x^2 + 1 \quad \text{mod } x^5
\end{aligned}
$$

i.e., $\Phi_{12}(x) = x^4 - x^2 + 1$. *This result follows because the denominator is a unit in the ring of polynomials modulo* x^t.

We also notice that

$$\Phi_n(0) = \begin{cases} -1, & n = 1 \\ 1, & n > 1. \end{cases}$$

It is remarked that the factorization in (6.6) is true over any field since the coefficients of $\Phi_d(x)$ are integers. Over a field of characteristic p the coefficients of $\Phi_d(x)$ are the integers modulo p, i.e., the coefficients are the elements of the prime field $\mathrm{GF}(p)$.

Returning to Example 6.3, each binomial factor $x^d - 1$ can be written as an expression containing only cyclotomic polynomials, i.e.,

$$
\begin{aligned}
V_{10}(x) &= \frac{(x-1)(x^{1023}-1)}{(x^3-1)(x^{31}-1)} = \frac{\Phi_1\Phi_1\Phi_3\Phi_{11}\Phi_{31}\Phi_{33}\Phi_{93}\Phi_{341}\Phi_{1023}}{\Phi_1\Phi_3\Phi_1\Phi_{31}} \\
&= \Phi_{11}\Phi_{33}\Phi_{93}\Phi_{341}\Phi_{1023},
\end{aligned}
$$

where, for short, $\Phi_s(x)$ is denoted simply as Φ_s. Since

$$
\begin{aligned}
\phi(11) &= \deg \Phi_{11} &= 10 \\
\phi(33) &= \deg \Phi_{33} &= 20 \\
\phi(93) &= \deg \Phi_{93} &= 60 \\
\phi(341) &= \deg \Phi_{341} &= 300 \\
\phi(1023) &= \deg \Phi_{1023} &= 600
\end{aligned}
$$

and $V_{10}(x)$ is the product of distinct monic irreducible polynomials of degree 10, it is concluded that over $\mathrm{GF}(2)$:

(a) $\Phi_{11}(x)$ is irreducible of degree 10

(b) $\Phi_{33}(x)$ is the product of two distinct irreducible degree 10 polynomials

(c) $\Phi_{93}(x)$ is the product of six distinct irreducible degree 10 polynomials

(d) $\Phi_{341}(x)$ is the product of 30 distinct irreducible degree 10 polynomials

(e) $\Phi_{1023}(x)$ is the product of 60 distinct irreducible degree 10 polynomials.

It is immediate to obtain

$$
\Phi_{11}(x) = \frac{x^{11}-1}{x-1} = x^{10} + x^9 + x^8 + x^7 + x^6 + x^5 + x^4 + x^3 + x^2 + x + 1,
$$

however, one needs to learn more about the factorization of cyclotomic polynomials in order to factor $\Phi_{33}, \Phi_{93}, \Phi_{341}$, and Φ_{1023}. A useful partial answer to this factorization problem is provided by the following theorem.

THEOREM 6.4 *If p is a prime and $\gcd(n,p) = 1$, then for $k \geq 1$ it follows that*

(1)
$$\Phi_{np^k}(x) = \Phi_{np}\left(x^{p^{k-1}}\right). \tag{6.9}$$

(2) $\Phi_{np^k}(x) = \dfrac{\Phi_n(x^{p^k})}{\Phi_n(x^{p^{k-1}})}.$

(3) $\Phi_{np^k}(x) = [\Phi_n(x)]^{(p^k - p^{k-1})}$ *over* $\mathrm{GF}(p^m)$.

Proof: The alternative form of the multiplicative Moebius inversion formula will now be used:

$$b(n) = \prod_{d:d|n} [a(n/d)]^{\mu(d)}$$

with $b(n) = \Phi_n(x)$ and $a(d) = x^d - 1$.

(1)
$$\Phi_{np^k}(x) = \prod_{d:d|np^k} \left[x^{np^k/d} - 1\right]^{\mu(d)} = \prod_{d:d|np} \left[x^{np^k/d} - 1\right]^{\mu(d)}$$

because $\mu(d) = 0$ if $p^u|d$, $u \geq 2$. Therefore it follows that

$$\prod_{d:d|np} \left[x^{np^k/d} - 1\right]^{\mu(d)} = \Phi_{np}(x^{p^{k-1}}).$$

(2) Expression (6.9) is now used and the values of d, $d|np$, are separated into two sets and the contribution of each set to $\Phi_{np^k}(x) = \Phi_{np}(x^{p^{k-1}})$ is calculated.

(a) Consider d, $d|np$, and $p|d$, $d = pd'$

$$\prod_{d':d'|n} \left[x^{np^{k-1}/d'} - 1\right]^{\mu(pd')} = \prod_{d':d'|n} \left[x^{np^{k-1}/d'} - 1\right]^{-\mu(d')}$$

$$= \left[\Phi_n(x^{p^{k-1}})\right]^{-1}.$$

(b) Consider d, $d|np$, and $p \nmid d$

$$\prod_{d:d|n} \left[x^{np^k/d} - 1\right]^{\mu(d)} = \Phi_n(x^{p^k})$$

and thus,

$$\Phi_{np}(x^{p^{k-1}}) = \frac{\Phi_n(x^{p^k})}{\Phi_n(x^{p^{k-1}})}.$$

(3) In a field of characteristic p it follows that $f(x^{p^r}) = [f(x)]^{p^r}, r \geq 1$, thus:

$$\Phi_{np}(x^{p^{k-1}}) = \frac{\Phi_n(x^{p^k})}{\Phi_n(x^{p^{k-1}})} = [\Phi_n(x)]^{(p^k - p^{k-1})}.$$

□

EXAMPLE 6.5 *The results of Theorem 6.4 will now be applied to $\Phi_{63}(x)$, noticing that $63 = 7.3^2$.*

(a) $\Phi_{7.3^2}(x) = \Phi_{7.3}(x^3) = \Phi_{21}(x^3)$.

(b) (b.1) d, $d|7.3$, and $3|d \Rightarrow d \in \{3, 21\}$,

(b.2) d, $d|7.3$, and $3 \nmid d \Rightarrow d \in \{1, 7\}$

$$
\begin{aligned}
\Phi_{21}(x^3) &= \frac{\Phi_7(x^9)}{\Phi_7(x^3)} = \frac{(x^{63} - 1)(x^3 - 1)}{(x^9 - 1)(x^{21} - 1)} \\
&= \frac{x^{42} + x^{21} + 1}{x^6 + x^3 + 1} \\
&= x^{36} + x^{33} + x^{27} + x^{24} + x^{18} + x^{12} + x^9 + x^3 + 1,
\end{aligned}
$$

which is the product of six irreducible polynomials of degree 6.

(c) *For a field of characteristic 3 it follows that*

$$\Phi_{21}(x^3) = [\Phi_7(x)]^{(9-3)} = [\Phi_7(x)]^6.$$

It is a property of finite fields that for a given n, over $\mathrm{GF}(q)$, $q = p^r$, $\gcd(n, p) = 1$, there is a least integer m such that $q^m \equiv 1$ modulo n. This value of m is called the order of q modulo n.

It was shown in Chapter 5 that there exists a finite field $\mathrm{GF}(q^m)$ with an element α of order n. Then, for every divisor d of n, similarly to (6.5), it also follows that

$$\Phi_d(x) = \prod_{s=0}^{n-1} (x - \alpha^s), \ \gcd(s, n) = n/d,$$

which can be equivalently expressed as

$$\Phi_d(x) = \prod_{\beta} (x - \beta), \ \mathrm{ord}(\beta) = d,$$

i.e., over $GF(q^m)$, $\Phi_d(x)$ factors completely into linear factors. By making use of minimal polynomials the factorization of $\Phi_d(x)$ over $GF(q)$ will now be considered.

Since the minimal polynomial of α has m conjugates $\alpha, \alpha^q, \ldots, \alpha^{q^{m-1}}$, it is concluded that $\Phi_n(x)$ has at least one irreducible factor of degree m, which is the minimal polynomial of α. By the definition of a cyclotomic polynomial, each one of the other roots of $\Phi_n(x)$ has order n and also m conjugates. These results are now stated as a theorem.

THEOREM 6.6 *Over* $GF(q)$, $q = p^r$, $p \nmid n$, *the cyclotomic polynomial* $\Phi_n(x)$ *consists of the product of* $\phi(n)/m$ *irreducible factors of degree* m, *where* m *is the order of* q *modulo* n.

A practical procedure to factor polynomials over $GF(q)$ which follows from the application of Theorem 6.7 is presented next (Berlekamp 1968, p.146).

THEOREM 6.7 *Suppose* $f(x)$ *is a monic polynomial of degree* n *with coefficients in* $GF(q)$. *If* $h(x) \in GF(q)[x]$ *is such that*

$$[h(x)]^q \equiv h(x) \mod f(x), \tag{6.10}$$

then

$$f(x) = \prod_{u \in GF(q)} \gcd[f(x), h(x) - u]. \tag{6.11}$$

Proof: From Theorem 5.4 one may infer that

$$z^q - z = \prod_{u \in GF(q)} (z - u),$$

and thus it follows that for any polynomial $h(x)$

$$[h(x)]^q - h(x) = \prod_{u \in GF(q)} (h(x) - u).$$

The hypothesis can thus be expressed as $f(x) \Big| \prod_{u \in GF(q)} (h(x) - u)$, which implies

$$f(x) \Big| \prod_{u \in GF(q)} \gcd[f(x), h(x) - u], \tag{6.12}$$

however, for each $u \in GF(q)$ it follows that

$$\gcd[f(x), h(x) - u] \Big| f(x). \tag{6.13}$$

Furthermore, since the difference $[(h(x) - u_1) - (h(x) - u_2)]$ is a scalar $u_2 - u_1 \in \mathrm{GF}(q)$, it follows that $h(x) - u_1$ and $h(x) - u_2$ are relatively prime for $u_1 \neq u_2$. Thus, $\gcd[f(x), h(x) - u_1]$ and $\gcd[f(x), h(x) - u_2]$ are relatively prime and from (6.13) it follows that

$$\prod_{u \in \mathrm{GF}(q)} \gcd[f(x), h(x) - u] \;\Big|\; f(x). \tag{6.14}$$

By comparing (6.12) and (6.14) it is concluded that

$$f(x) = \prod_{u \in \mathrm{GF}(q)} \gcd[f(x), h(x) - u].$$

$\square$

REMARKS

(1) If, for some $u \in \mathrm{GF}(q)$, it is true that $h(x) \equiv u$ modulo $f(x)$ then the factorization is said to be trivial as one of the factors is $f(x)$ and all other factors are equal to 1.

(2) In general $\gcd[f(x), h(x) - u]$ may be reducible in $\mathrm{GF}(q)[x]$. By using all the $h(x)$ as in Theorem 6.7 the factorization of $f(x)$ is obtained.

The set of polynomials reduced modulo $f(x)$ can be seen to form an n dimensional vector space $V(f)$ over $\mathrm{GF}(q)$. A suitable basis for $V(f)$ is provided by $\{1, x, x^2, \ldots, x^{n-1}\}$.

DEFINITION 6.8 *The polynomials $h(x)$ which allow the factorization of $f(x)$ are called f-reducing polynomials.*

Let $R(f)$ denote the subset of $V(f)$ containing the f-reducing polynomials. The set $R(f)$ constitutes a subspace of $V(f)$ because if $h_1(x) \in R(f)$, $h_2(x) \in R(f)$ and $c_1 \in \mathrm{GF}(q)$, $c_2 \in \mathrm{GF}(q)$, where both c_1 and c_2 are nonzero, it follows that

$$\begin{aligned}
h(x) &= c_1 h_1(x) + c_2 h_2(x) \\
h^q(x) &= [c_1 h_1(x) + c_2 h_2(x)]^q = c_1^q h_1^q(x) + c_2^q h_2^q(x) \\
&= c_1 h_1^q(x) + c_2 h_2^q(x) \equiv c_1 h_1(x) + c_2 h_2(x) \quad \mathrm{mod}\ f(x).
\end{aligned}$$

THEOREM 6.9 *The dimension of $R(f)$ equals the number of distinct irreducible factors of $f(x)$, denoted by k.*

The polynomial $x^n - 1$, with $\gcd(n, q) = 1$, allows the Berlekamp algorithm to be simplified as follows.

THEOREM 6.10 *A polynomial* $h(x) = \sum_{i=0}^{n-1} h_i x^i$ *satisfies the congruence* $h^q(x) \equiv h(x)$ *modulo* $(x^n - 1)$ *if and only if* $h_{iq} = h_i$ *for all subscripts* $i = 0, 1, \ldots, n-1$ *reduced modulo* n.

Proof:

$$h^q(x) = \left(\sum_{i=0}^{n-1} h_i x^i \right)^q = \sum_{i=0}^{n-1} h_i^q x^{iq} \equiv \sum_{i=0}^{n-1} h_i x^{iq} \mod (x^n - 1),$$

and the theorem follows when the coefficients of x^{iq} *on both sides of the congruence are compared.* $\square$

EXAMPLE 6.11 *Let* $f(x) = x^4 + x^3 + x^2 + 1$ *and let* $q = 2$. *If* $h(x) = h_3 x^3 + h_2 x^2 + h_1 x + h_0$ *is considered and the condition* $h^2(x) \equiv h(x)$ *modulo* $f(x)$ *is applied then it follows that*

$$h_3 x^6 + h_2 x^4 + h_1 x^2 + h_0 \equiv h_3 x^3 + h_2 x^2 + h_1 x + h_0 \tag{6.15}$$

modulo $x^4 + x^3 + x^2 + 1$, *where* $h_i^2 = h_i$ *since* $h_i \in \{0, 1\}$, $0 \le i \le 3$. *However,*

$$x^4 \equiv x^3 + x^2 + 1 \mod (x^4 + x^3 + x^2 + 1)$$
$$x^6 \equiv x^3 + x^2 + x \mod (x^4 + x^3 + x^2 + 1).$$

Representing the polynomial $h_0 + h_1 x + h_2 x^2 + h_3 x^3$ *by the column vector* $(h_0, h_1, h_2, h_3)^T$, *the congruence* (6.15) *can be expressed as*

$$h_0 \begin{bmatrix} 1 \\ 0 \\ 0 \\ 0 \end{bmatrix} + h_1 \begin{bmatrix} 0 \\ 0 \\ 1 \\ 0 \end{bmatrix} + h_2 \begin{bmatrix} 1 \\ 0 \\ 1 \\ 1 \end{bmatrix} + h_3 \begin{bmatrix} 0 \\ 1 \\ 1 \\ 1 \end{bmatrix}$$

$$\equiv h_0 \begin{bmatrix} 1 \\ 0 \\ 0 \\ 0 \end{bmatrix} + h_1 \begin{bmatrix} 0 \\ 1 \\ 0 \\ 0 \end{bmatrix} + h_2 \begin{bmatrix} 0 \\ 0 \\ 1 \\ 0 \end{bmatrix} + h_3 \begin{bmatrix} 0 \\ 0 \\ 0 \\ 1 \end{bmatrix}.$$

Using the binary column vectors on the left-hand side to form a 4×4 *matrix and similarly for the binary column vectors on the right-hand side, it is desired to find* $h = [h_0, h_1, h_2, h_3]$ *which is in the null space of the following matrix* **B**:

$$\mathbf{B} = \begin{bmatrix} 1 & 0 & 1 & 0 \\ 0 & 0 & 0 & 1 \\ 0 & 1 & 1 & 1 \\ 0 & 0 & 1 & 1 \end{bmatrix} - \begin{bmatrix} 1 & 0 & 0 & 0 \\ 0 & 1 & 0 & 0 \\ 0 & 0 & 1 & 0 \\ 0 & 0 & 0 & 1 \end{bmatrix} = \begin{bmatrix} 0 & 0 & 1 & 0 \\ 0 & 1 & 0 & 1 \\ 0 & 1 & 0 & 1 \\ 0 & 0 & 1 & 0 \end{bmatrix}$$

or

$$\begin{cases} h_2 & = & 0 \\ h_1 + h_3 & = & 0 \end{cases}$$

and the solutions for (h_0, h_1, h_2, h_3) *are* $(0,0,0,0)$, $(1,0,0,0)$, $(0,1,0,1)$ *and* $(1,1,0,1)$. *The dimension of the subspace of the* h's *is two, which means that* $f(x)$ *has two distinct irreducible factors given by* $\gcd(f(x), x^3 + x + 1) = x^3 + x + 1$, $\gcd(f(x), x^3 + x) = x + 1$.

6.3   Canonical factorization

Let $f(x)$ be any polynomial of positive degree, $f(x) \in \mathrm{GF}(q)[x]$. By the canonical factorization of $f(x)$ over $\mathrm{GF}(q)[x]$ it is meant to express $f(x)$ as

$$f(x) = A_0 f_1^{e_1}(x) \cdot f_2^{e_2}(x) \ldots f_k^{e_k}(x),$$

where $f_i(x)$, $1 \le i \le k$, are distinct monic irreducible polynomials in $\mathrm{GF}(q)[x]$ and e_i, $1 \le i \le k$, are positive integers, and $A_0 \ne 0$ is an element of $\mathrm{GF}(q)$. It can be shown that any polynomial $f(x) \in \mathrm{GF}(q)[x]$ of positive degree has a canonical factorization over $\mathrm{GF}(q)[x]$.

6.4   Eliminating repeated factors

The problem of factoring $f(x)$ can be simplified by considering only the factorization of polynomials without repeated factors as follows. Calculate $s(x) = \gcd(f(x), f'(x))$, the greatest common divisor between $f(x)$ and its derivative $f'(x)$.

(a) If $s(x) = 1$ then $f(x)$ has no repeated factor.

(b) If $s(x) = f(x)$ then $f'(x) = 0$ and hence $f(x) = [g(x)]^{p^r}$, where p is the characteristic of $\mathrm{GF}(q)$ and r is a positive integer. If desired, the procedure can be applied to $g(x)$.

(c) If $s(x) \ne 1$ and $s(x) \ne f(x)$, then $s(x)$ is a nontrivial factor of $f(x)$ and $f(x)/s(x)$ has no repeated factor. In case $s(x)$ has repeated factors keep using the reduction process as long as necessary.

6.5   Irreducibility of $\Phi_n(x)$ over $\mathrm{GF}(q)$

From Theorem 6.6 it is known that $\Phi_n(x)$ is irreducible over $\mathrm{GF}(q)$ if and only if $\phi(n)/m = 1$, where $q^m \equiv 1$ modulo n and $q^i \ne 1$, $0 < i < m$. The set of residues modulo n which are relatively prime to n form a multiplicative group of order $\phi(n)$. In the case of $q^{\phi(n)} \equiv 1$ modulo n a cyclic group is obtained with q as its generator, i.e., q is a primitive root modulo n. In number theory it is shown that the only values of n

for which the group of residues is cyclic modulo n are $n = 1, 2, 4, p^u, 2p^u$ where p is an odd prime.

EXAMPLE 6.12 *Consider* $\Phi_8(x) = x^4 + 1$. *Then* $\phi(8) = 4$, *but there is no element of order 4 modulo 8 and thus* $\Phi_8(x)$ *is reducible modulo p for every p. On the other hand, for the important case of $q = 2$, $\Phi_n(x)$ is reducible over* GF(2) *if and only if 2 is a primitive root modulo n, i.e., if and only if 2 has order $\phi(n)$ modulo n. This restriction conditions the values of n to be* $3, 5, 9, 11, 13, 19, \ldots$ *and produces irreducible polynomials of degrees* $2, 4, 6, 10, 12, 18, \ldots$.

EXAMPLE 6.13 *Factor* $f(x) = x^5 + x^4 + 1$ *over* GF(2) *using Berlekamp's algorithm.*

Solution: *Let* $h(x) = h_0 + h_1 x + h_2 x^2 + h_3 x^3 + h_4 x^4$. *Proceeding as in Example 6.11 it follows that*

$$
\mathbf{B} = \begin{bmatrix}
0 & 0 & 0 & 1 & 1 \\
0 & 1 & 0 & 1 & 1 \\
0 & 1 & 1 & 0 & 1 \\
0 & 0 & 0 & 1 & 1 \\
0 & 0 & 1 & 1 & 0
\end{bmatrix}.
$$

Solving the system of linear equations

$$(h_0, h_1, h_2, h_3, h_4)\mathbf{B}^\mathrm{T} = (0, 0, 0, 0, 0),$$

the solutions obtained for $(h_0, h_1, h_2, h_3, h_4)$ *are the following* $(0, 0, 0, 0, 0)$, $(1, 0, 0, 0, 0)$, $(0, 0, 1, 1, 1)$, *and* $(1, 0, 1, 1, 1)$. *For these solutions it follows that*

(a) $h(x) = x^4 + x^3 + x^2$ *and that* $\gcd(f(x), x^4 + x^3 + x^2) = x^2 + x + 1$.

(b) $h(x) = x^4 + x^3 + x^2 + 1$ *and that* $\gcd(f(x), x^4 + x^3 + x^2 + 1) = x^3 + x + 1$.

Thus $x^5 + x^4 + 1 = (x^3 + x + 1)(x^2 + x + 1)$ *is the answer because* $x^3 + x + 1$ *and* $x^2 + x + 1$ *are both known to be irreducible over* GF(2).

EXAMPLE 6.14 *Factor* $f(x) = x^6 + x^2 + 1$ *over* GF(2) *using Berlekamp's algorithm.*

Solution: *Let* $h(x) = h_0 + h_1 x + h_2 x^2 + h_3 x^3 + h_4 x^4 + h_5 x^5$ *and apply Theorem 6.7, i.e.,* $h^2(x) \equiv h(x) \mod f(x)$, *where* $f(x) = x^6 + x^2 + 1$.

It then follows that

$$
h_0 \begin{bmatrix} 1 \\ 0 \\ 0 \\ 0 \\ 0 \\ 0 \end{bmatrix} + h_1 \begin{bmatrix} 0 \\ 0 \\ 1 \\ 0 \\ 0 \\ 0 \end{bmatrix} + h_2 \begin{bmatrix} 0 \\ 0 \\ 0 \\ 0 \\ 1 \\ 0 \end{bmatrix} + h_3 \begin{bmatrix} 1 \\ 0 \\ 1 \\ 0 \\ 0 \\ 0 \end{bmatrix} + h_4 \begin{bmatrix} 0 \\ 0 \\ 1 \\ 0 \\ 1 \\ 0 \end{bmatrix} + h_5 \begin{bmatrix} 1 \\ 0 \\ 1 \\ 0 \\ 1 \\ 0 \end{bmatrix} \equiv
$$

$$
h_0 \begin{bmatrix} 1 \\ 0 \\ 0 \\ 0 \\ 0 \\ 0 \end{bmatrix} + h_1 \begin{bmatrix} 0 \\ 1 \\ 0 \\ 0 \\ 0 \\ 0 \end{bmatrix} + h_2 \begin{bmatrix} 0 \\ 0 \\ 1 \\ 0 \\ 0 \\ 0 \end{bmatrix} + h_3 \begin{bmatrix} 0 \\ 0 \\ 0 \\ 1 \\ 0 \\ 0 \end{bmatrix} + h_4 \begin{bmatrix} 0 \\ 0 \\ 0 \\ 0 \\ 1 \\ 0 \end{bmatrix} + h_5 \begin{bmatrix} 0 \\ 0 \\ 0 \\ 0 \\ 0 \\ 1 \end{bmatrix}
$$

modulo $x^6 + x^2 + 1$, *or* $(h_0, h_1, h_2, h_3, h_4, h_5)\mathbf{B}^{\mathrm{T}} = (0,0,0,0,0,0)$ *with*

$$
\mathbf{B} = \begin{bmatrix} 0 & 0 & 0 & 1 & 0 & 1 \\ 0 & 1 & 0 & 0 & 0 & 0 \\ 0 & 1 & 1 & 1 & 1 & 1 \\ 0 & 0 & 0 & 1 & 0 & 0 \\ 0 & 0 & 1 & 0 & 0 & 1 \\ 0 & 0 & 0 & 0 & 0 & 1 \end{bmatrix}.
$$

The only solutions are $\mathbf{h} = [0,0,0,0,0,0]$ *and* $\mathbf{h} = [1,0,0,0,0,0]$. *Since* $f'(x) = 0$ *over* $\mathrm{GF}(2)$ *it follows that* $\gcd(f(x), f'(x)) = f(x)$ *and that* $x^6 + x^2 + 1 = (x^3 + x + 1)^2$, *where* $x^3 + x + 1$ *is irreducible over* $\mathrm{GF}(2)$.

EXAMPLE 6.15 *Factor* $\Phi_{15}(x)$ *over* $\mathrm{GF}(2)$.

Solution: *Since the order of 2 modulo 15 is 4, i.e., 4 is the least positive integer for which* $2^4 \equiv 1$ *modulo 15, and* $\phi(15) = 8$, *it follows that* $\Phi_{15}(x)$ *factors over* $\mathrm{GF}(2)$ *as the product of* $\phi(15)/4 = 8/4 = 2$ *irreducible polynomials of degree 4. Let* α *denote an element of order 15 in* $\mathrm{GF}(2^4)$. *The two irreducible polynomials are obtained from*

$$
\begin{aligned}
f_1(x) &= (x - \alpha)(x - \alpha^2)(x - \alpha^4)(x - \alpha^8) \\
f_2(x) &= (x - \alpha^7)(x - \alpha^{14})(x - \alpha^{13})(x - \alpha^{11}),
\end{aligned}
$$

with $\alpha^4 = \alpha + 1$ *it follows that* $f_1(x) = x^4 + x + 1$ *and* $f_2(x) = x^4 + x^3 + 1$.

EXAMPLE 6.16 *Factor* $\Phi_{20}(x)$ *over* $\mathrm{GF}(3)$.

Solution: *Notice that* $\phi(20) = 8$ *and that the order of 3 modulo 20 is 4, i.e., 4 is the least positive integer for which* $3^4 \equiv 1$ *modulo 20. Thus* $\phi(20)/4 = 8/4 = 2$ *and* $\Phi_{20}(x)$ *factors over* $\mathrm{GF}(3)$ *as the product of two* *degree 4 irreducible polynomials. Let α denote an element of order 20 in* $\mathrm{GF}(3^4)$. *The two irreducible polynomials are obtained from*

$$
\begin{array}{rcl}
f_1(x) & = & (x - \alpha)(x - \alpha^3)(x - \alpha^9)(x - \alpha^7) \\
f_2(x) & = & (x - \alpha^{11})(x - \alpha^{13})(x - \alpha^{19})(x - \alpha^{17}),
\end{array}
$$

EXAMPLE 6.17 *Factor $x^4 + 1$ over* $\mathrm{GF}(3)$.

Solution:

(a) *Since $\frac{d}{dx}(x^4 + 1) = x^3$ it follows that $\gcd(x^4 + 1, x^3) = 1$. Thus $x^4 + 1$ has no repeated factors.*

(b) *Let $h(x) = h_0 + h_1 x + h_2 x^2 + h_3 x^3$. The following $\mathbf{B}^{\mathrm{T}}$ is obtained*

$$
\mathbf{B}^{\mathrm{T}} = \begin{bmatrix}
0 & 0 & 0 & 0 \\
0 & 2 & 0 & 1 \\
0 & 0 & 1 & 0 \\
0 & 1 & 0 & 2
\end{bmatrix}.
$$

By replacing column 2 in $\mathbf{B}^{\mathrm{T}}$ by the sum of columns 2 and 4 the following matrix results:

$$
\begin{bmatrix}
0 & 0 & 0 & 0 \\
0 & 0 & 0 & 1 \\
0 & 0 & 1 & 0 \\
0 & 0 & 0 & 2
\end{bmatrix}.
$$

It follows that $\mathbf{B}$ has rank 2 and thus $x^4 + 1$ has $k = 4 - 2 = 2$ irreducible factors. The following system of linear equations results

$$
\begin{cases}
h_2 & = & 0 \\
h_1 + 2h_3 & = & 0
\end{cases}
$$

which has $h_1(x) = 1$ and $h_2(x) = x^3 + x$ as solutions. Finally,

$$
\begin{array}{rcl}
\gcd(x^4 + 1, h_2(x) + 1) & = & x^2 + x + 2 \\
\gcd(x^4 + 1, h_2(x) + 2) & = & x^2 + 2x + 2.
\end{array}
$$

EXAMPLE 6.18 *Determine the number of distinct monic irreducible factors of $x^4 + 1$ in $\mathrm{GF}(p)[x]$, for all odd primes p. Hint:*

$$
\Phi_8(x) = \frac{x^8 - 1}{x^4 - 1} = x^4 + 1,
$$

and use Theorem 6.6.

Solution: *From Theorem 6.6 it follows that the number of distinct monic irreducible factors of $\Phi_8(x)$ over $\mathrm{GF}(p)$ is given by $\phi(8)/m = 4/m$,* where m is the order of p modulo 8. It is immediate to show that $m \leq 2$ if p is any odd prime by noticing that $p^2 \equiv 1$ modulo 8, because if p is written as $p = 2r + 1$ then

$$p^2 - 1 = (p - 1)(p + 1) = 2r(2r + 2) = 4r(r + 1)$$

and it follows that $p^2 - 1 \equiv 0$ modulo 8, whether r is even or odd. It thus follows that $\phi(8)/m = 4/2 = 2$, i.e., $x^4 + 1$ factors as the product of two polynomials of degree 2 over $GF(p)[x]$, if $m = 2$. Notice that if $p \equiv 1$ modulo 8, e.g., $p = 17$, then $m = 1$ and $x^4 + 1$ factors as the product of $4/1 = 4$ monic linear factors.

EXAMPLE 6.19 *Factor* $x^3 + 4x^2 + x - 1$ *over* $GF(5)[x]$ *using Berlekamp's algorithm.*

Solution: Let $h(x) = h_2 x^2 + h_1 x + h_0$. Next compute the residue of $h^5(x)$ modulo $f(x)$, i.e., compute x^5 modulo $f(x)$ and compute x^{10} modulo $f(x)$. However, $x^5 \equiv x$ modulo $f(x)$. It thus follows immediately that $h(x) = x$ and

$$\begin{aligned}
\gcd(f(x), x - 1) &= x - 1 \\
\gcd(f(x), x - 2) &= x - 2 \\
\gcd(f(x), x - 3) &= x - 3.
\end{aligned}$$

6.6   Problems with solutions

(1) If n is a positive integer then prove that

$$\Pi_{d:\ d|n}(-1)^{\mu(d)} = \begin{cases} -1, & \text{if } n = 1 \\ 1, & \text{if } n > 1. \end{cases}$$

Solution: Obviously, in $n = 1$ it follows that

$$\prod_{d:d|1}(-1)^{\mu(d)} = (-1)^1 = -1.$$

Let $n > 1$ and suppose the prime factor decomposition of n is given by $p_1^{e_1} p_2^{e_2} \cdots p_s^{e_s}$, where the p_i denote prime numbers and the e_i denote positive integers, $1 \leq i \leq s$. For computing $\mu(d)$, when $d|n$, only those d's of the form $p_1^{i_1} p_2^{i_2} \cdots p_s^{i_s}$ need to be considered, where $i_j \in \{0, 1\}, 1 \leq j \leq s$. The other divisors of n, i.e., those with at least one $i_j \geq 2$, lead to $\mu(d) = 0$. Therefore, 2^s divisors of n are being considered. Among those divisors, half of them contain an odd number of factors and the other half contains an even number of factors. Therefore, it follows that

$$\prod_{d|n}(-1)^{\mu(d)} = (-1)^{2^s-1}/(-1)^{2^s-1} = 1, \text{ if } n > 1.$$

(2) If n is odd then prove that $\Phi_{2n}(x) = \Phi_n(-x)$, $n > 1$.

Solution: By Theorem 6.4, part b, for n odd, it follows that

$$\Phi_{2n}(x) = \frac{\Phi_n(x^2)}{\Phi_n(x)}$$

$$\Phi_n(x^2) = \prod_{d|n}(x^{2n/d}-1)^{\mu(d)} = \prod_{d|n}[(x^{n/d}-1)(x^{n/d}+1)]^{\mu(d)}$$

$$= \Phi_n(x)\prod_{d|n}(x^{n/d}+1)^{\mu(d)} \Rightarrow \Phi_{2n}(x) = \prod_{d|n}(x^{n/d}+1)^{\mu(d)}$$

$$\Phi_n(-x) = \prod_{d|n}[(-x)^{n/d}-1]^{\mu(d)}$$

$$= \prod_{d|n}(x^{n/d}+1)^{\mu(d)}\prod_{d|n}(-1)^{\mu(d)}, n \text{ odd}.$$

However, by Problem 1, $\prod_{d|n}(-1)^{\mu(d)} = 1$, if $n > 1$, and it follows that $\Phi_{2n}(x) = \Phi_n(-x)$, n odd, $n > 1$.

(3) Show that cyclotomic polynomials are self-reciprocal polynomials.

Solution: If $f_1(x)$ and $f_2(x)$ are reciprocal polynomials, it follows that if $f_1(\alpha) = 0$ then $f_2(1/\alpha) = 0$. Suppose α is an element of order n in $GF(q^m)$. Then by the definition of $\Phi_n(x)$ it follows that $\Phi_n(\alpha) = 0$. Let $\beta = 1/\alpha$. Then, β also has order n in $GF(q^m)$ and thus $\Phi_n(\beta) = 0$.

(4) Prove that the constant term of a cyclotomic polynomial $\Phi_n(x)$ is -1 if $n - 1$, and $+1$ if $n > 1$.

Solution: Write $\Phi_n(x)$ as $\Phi_n(x) = \prod_{d|n}(x^{n/d}-1)^{\mu(d)}$ and by making $x = 0$ it follows that $\Phi_n(0) = \prod_{d|n}(-1)^{\mu(d)}$, i.e., precisely the same expression obtained in Problem 1, which solves this problem since $\Phi_n(0)$ represents the constant term in $\Phi_n(x)$.

(5) Calculate the following cyclotomic polynomials.

(a) $\Phi_{24}(x)$.

(b) $\Phi_{40}(x)$.

(c) $\Phi_{60}(x)$.

Solution:

(a) By (6.7) and reducing $\mod x^t$, for $t = \phi(24) + 1 = 9$, it follows that

$$\Phi_{24}(x) = \frac{\prod_{d|3}(x^{3.8/d} - 1)^{\mu(d)}}{\prod_{d|3}(x^{3.4/d} - 1)^{\mu(d)}} = \frac{(x^{24} - 1)(x^4 - 1)}{(x^{12} - 1)(x^8 - 1)}$$

$$= \frac{(-1)(x^4 - 1)}{(-1)(x^8 - 1)} \mod 9 = \frac{(x^4 - 1)(x^8 + 1)}{(x^8 - 1)(x^8 + 1)}$$

$$= \frac{(x^4 - 1)(x^8 + 1)}{-1} \mod x^9 = x^8 - x^4 + 1.$$

(b) By (6.7) it follows that

$$\Phi_{40}(x) = \frac{(x^{40} - 1)(x^4 - 1)}{(x^8 - 1)(x^{20} - 1)} = \frac{x^{20} + 1}{x^4 + 1}$$

$$= x^{16} + x^{12} + x^8 + x^4 + 1.$$

(c)

$$\Phi_{60}(x) = \frac{(x^{60} - 1)(x^4 - 1)}{(x^{20} - 1)(x^{12} - 1)} \frac{(x^{10} - 1)(x^6 - 1)}{(x^{30} - 1)(x^2 - 1)}$$

$$= \frac{(x^{30} + 1)(x^2 + 1)}{(x^{10} + 1)(x^6 + 1)}$$

$$= x^{16} + x^{14} - x^{10} - x^8 - x^6 + x^2 + 1.$$

(6) Calculate the canonical factorization of the following cyclotomic polynomials, over the given finite field.

(a) $\Phi_{17}(x)$, over $GF(2)$

(b) $\Phi_{11}(x)$, over $GF(3)$

(c) $\Phi_{13}(x)$, over $GF(5)$.

Solution:

(a) The order of 2 $\mod 17$ is 8, thus $\phi(17)/8 = 16/8 = 2$ and it follows that $\Phi_{17}(x)$ factors as the product of two irreducible polynomials of degree 8. Applying Theorem 6.7 with 8 as the order of 2 modulo 17, it follows that $\Phi_{17}(x) = (x^{17} - 1)/(x - 1)$, and that

$$\gcd(x^{17} - 1, x + x^2 + x^4 + x^8 + x^9 + x^{13} + x^{15} + x^{16})$$
$$= x^9 + x^8 + x^6 + x^3 + x + 1$$
$$= (x + 1)(x^8 + x^5 + x^4 + x^3 + 1).$$

and

$$\gcd(x^{17} - 1, 1 + x + x^2 + x^4 + x^8 + x^9 + x^{13} + x^{15} + x^{16})$$
$$= x^8 + x^7 + x^6 + x^4 + x^2 + x + 1.$$

Therefore,

$$\phi_{17}(x) = (x^8 + x^5 + x^4 + x^3 + 1)(x^8 + x^7 + x^6 + x^4 + x^2 + x + 1).$$

(b) The order of 3 mod 11 is 5 and $\Phi_{11}(x) = (x^{11} - 1)/(x - 1)$. It follows that

$$\gcd(x^{11} - 1, x + x^3 + x^4 + x^5 + x^9) = x^5 + 2x^3 + x^2 + 2x + 2,$$

and its reciprocal-root polynomial is $x^5 + x^4 + 2x^3 + x^2 + 2$. Thus,

$$\Phi_{11}(x) = (x^5 + 2x^3 + x^2 + 2x + 2)(x^5 + x^4 + 2x^3 + x^2 + 2).$$

(c) The order of 5 mod 13 is 4. Since $\phi(13)/4 = 3$, we need to find three degree 4 monic irreducible polynomials.

$$\begin{aligned}
\gcd(x^{13} - 1, x + x^5 + x^8 + x^{12}) &= 1 \\
\gcd(x^{13} - 1, 1 + x + x^5 + x^8 + x^{12}) &= x^5 + 3x^3 + 2x^2 + 4 \\
\gcd(x^{13} - 1, 2 + x + x^5 + x^8 + x^{12}) &= x^4 + 2x^3 + x^2 + 2x + 1 \\
\gcd(x^{13} - 1, 3 + x + x^5 + x^8 + x^{12}) &= x^4 + 3x^3 + 3x + 1 \\
\gcd(x^{13} - 1, 4 + x + x^5 + x^8 + x^{12}) &= 1,
\end{aligned}$$

where $x^5 + 3x^3 + 2x^2 + 4$ factors as $(x - 1)(x^4 + x^3 + 4x^2 + x + 1)$. Finally, $\Phi_{13}(x)$ is equal to

$$(x^4 + x^3 + 4x^2 + x + 1)(x^4 + 2x^3 + x^2 + 2x + 1)(x^4 + 3x^3 + 3x + 1).$$

(7) Determine the shape of the factorization of $\Phi_{360}(x)$ over GF(3) by using Theorem 6.4.

Solution: Consider the following factorization: $360 = 2^3 3^2 5$. It follows that

$$\Phi_{360}(x) = \frac{\Phi_{40}(x^9)}{\Phi_{40}(x^3)} = \frac{\Phi_5(x^{72})/\Phi_5(x^{36})}{\Phi_5(x^{24})/\Phi_5(x^{12})}$$

(8) Find the canonical factorization of $x^{15} - 1$ over $GF(4)$.

Solution: Cyclotomic cosets modulo 15 over $GF(4)$: (0), $(1,4)$, $(2,8)$, $(3,12)$, (5), $(6,9)$, $(7,13)$, (10), $(11,14)$.

$$
\begin{aligned}
\gcd(x^{15}+1, x^4+x) &= x^3+1 = (x+1)(x^2+x+1) \\
&= (x+1)(x+\alpha)(x+\alpha^2) \\
\gcd(x^{15}+1, x^4+x+1) &= x^4+x+1 \\
\gcd(x^{15}+1, x^4+x+\alpha^2) &= x^4+x+\alpha^2 \\
\gcd(x^{15}+1, x^{13}+x^7+\alpha) &= x^4+\alpha x^3+\alpha \\
\gcd(x^4+\alpha x^3+\alpha, x^4+x+1) &= x^2+x+\alpha \\
\gcd(x^4+\alpha x^3+\alpha, x^4+x+\alpha^2) &= x^2+\alpha^2 x+1.
\end{aligned}
$$

Continuing in this way we obtain the factorization of $x^{15}+1$ as the product of the following polynomials: $x+1, x+\alpha, x+\alpha^2, x^2+x+\alpha, x^2+\alpha^2 x+\alpha^2, x^2+x+\alpha^2, x^2+\alpha x+\alpha, x^2+\alpha^2 x+1, x^2+\alpha x+1$, obtained by using $\alpha^2 = \alpha+1$, $\alpha \in GF(4)$, and $\beta^2 = \beta+\alpha$, $\beta \in GF(16)$, and the following table for the powers of β.

0	$\beta^4 = \beta+1$	$\beta^9 = \alpha\beta+\alpha$	$\beta^{14} = \alpha^2\beta+\alpha^2$
1	$\beta^5 = \alpha$	$\beta^{10} = \alpha^2$	
β	$\beta^6 = \alpha\beta$	$\beta^{11} = \alpha^2\beta$	
$\beta^2 = \beta+\alpha$	$\beta^7 = \alpha^2\beta$	$\beta^{12} = \alpha^2\beta+1$	
$\beta^3 = \alpha^2\beta+\alpha$	$\beta^8 = \beta+\alpha^2$	$\beta^{13} = \alpha\beta+1$	

Chapter 7

CONSTRUCTING F-REDUCING POLYNOMIALS

7.1 Introduction

Let $f(x)$ be a monic polynomial of degree n without repeated factors. We recall from Definition 6.8 that f-reducing polynomials are those polynomials which allow the factorization of $f(x)$. In this chapter, we look into the interesting problem of factorization of polynomials over large finite fields using f-reducing polynomials.

Let $f_1(x), f_2(x), \ldots, f_k(x)$ be monic irreducible factors in the canonical factorization (see Section 6.3) of $f(x)$ in $\mathrm{GF}(q)[x]$, where $\deg(f_i(x)) = n_i$, $1 \leq i \leq k$. Let N be the least positive integer such that $x^{q^N} \equiv x$ mod $f(x)$. By Theorem 6.10 it follows that

$$T_i(x) = T(x^i) = \sum_{j=0}^{N-1} x^{iq^j}$$

are f-reducing polynomials. It is immediate to check that

$$T_i^q(x) \equiv T_i(x) \quad \mod f(x), \ 1 \leq i \leq n-1.$$

Whenever the order of $f(x)$ is known in advance, e.g., when $f(x) = x^n - 1$, or $f(x) = \Phi_n(x)$, it is convenient to employ the polynomial $R_i(x)$ defined as

$$R_i(x) = x^i + x^{iq} + x^{iq^2} + \ldots + x^{iq^{m_i-1}},$$

where, for each $i > 0$, m_i is the least positive integer such that

$$x^{iq^{m_i}} \equiv x^i \quad \mod f(x).$$

It is assumed that $f(0) \neq 0$. If $\mathrm{ord}(f(x)) = e$ then the condition on m_i is equivalent to

$$iq^{m_i} \equiv i \quad \mod e \text{ or } q^{m_i} \equiv 1 \quad \mod e/\gcd(e, i),$$

i.e., m_i is the multiplicative order of q modulo $e/\gcd(e, i)$. Notice that

$$R_i^q(x) \equiv R_i(x) \mod f(x)$$

and Theorem 6.7 thus applies.

7.2   Factoring polynomials over large finite fields

Obviously, the basic methods of factorization of polynomials over finite fields remain applicable for large finite fields. However, the amount of computation increases with the number of elements in $\mathrm{GF}(q)$. It is thus important to find ways to avoid computing gcd's which are equal to one because they lead to trivial results.

In the present context, to say that q is large means q is large with respect to n, the degree of $f(x)$, i.e., the polynomial to be factored. If $f(x)$ has k distinct monic irreducible factors, i.e., k prime polynomials, in $\mathrm{GF}(q)[x]$, then at most k gcd's will be different from 1. In the sequel the elements $c \in \mathrm{GF}(q)$ for which $\gcd(f(x), h(x) - c) \neq 1$ will be characterized. One technique makes use of the *resultant* of two polynomials, which is defined next.

7.2.1   Resultant

Let

$$f(x) = \sum_{i=0}^{n} a_i x^i \in \mathrm{GF}(q)[x]$$

and let

$$g(x) = \sum_{j=0}^{m} b_j x^j \in \mathrm{GF}(q)[x]$$

be two polynomials of formal degree n and m, respectively, where $n \in \mathbb{N}$ and $m \in \mathbb{N}$.

DEFINITION 7.1 *The resultant* $R = R(f(x), g(x))$ *of the polynomials* $f(x)$ *and* $g(x)$ *is defined by the following determinant of order* $n + m$

$$R = \left| \begin{array}{cccccccc} a_n & a_{n-1} & \cdots & \cdots & a_1 & a_0 & \cdots & 0 \\ 0 & a_n & \cdots & \cdots & a_2 & a_1 & \cdots & 0 \\ \vdots & & & & \vdots & & & \vdots \\ 0 & 0 & 0 & 0 & a_n & a_{n-1} & \cdots & a_0 \\ b_m & b_{m-1} & \cdots & b_0 & 0 & 0 & \cdots & 0 \\ 0 & b_m & \cdots & \cdots & b_0 & 0 & \cdots & 0 \\ \vdots & & & & \vdots & & & \vdots \\ 0 & 0 & \cdots & \cdots & b_m & b_{m-1} & \cdots & b_0 \end{array} \right| \begin{array}{l} \left. \begin{array}{l} \\ \\ \\ \end{array} \right\} m \text{ rows} \\ \left. \begin{array}{l} \\ \\ \\ \end{array} \right\} n \text{ rows} \end{array}$$

From the definition of resultant it is shown next that $R = 0$ if and only if $f(x)$ and $g(x)$ have a common factor. Notice that $R = 0$ if and only if the rows of the determinant R are linearly dependent, i.e., if and only if

$$c_1(x)f(x) + c_2(x)g(x) = 0,$$

where $\deg c_1(x) < m$ and $\deg c_2(x) < n$. Dividing $c_1(x)f(x) + c_2(x)g(x)$ throughout by $f(x)$ it is concluded that $g(x)$ must have a factor in common with $f(x)$, because $\deg c_2(x) < n$ and thus $c_2(x)$ is not divisible by $f(x)$. By using a similar argument the same conclusion will be reached if $c_1(x)f(x) + c_2(x)g(x)$ is divided through by $g(x)$.

Let $f(x) = a(x)F(x)$ and let $g(x) = a(x)G(x)$. It follows that

$$f(x)g(x) = a(x)G(x)f(x) = a(x)F(x)g(x), \quad a(x) \neq 0,$$

and

$$G(x)f(x) - F(x)g(x) = 0,$$

thus $c_1(x) = G(x)$ and $c_2(x) = -F(x)$.

7.2.2 Algorithm for factorization based on the resultant

Let $h(x)$ be an f-reducing polynomial and consider the resultant $R = R(f(x), h(x) - c)$. It has already been seen that $\gcd(f(x), h(x) - c) \neq 1$ if and only if $R(f(x), h(x) - c) = 0$. Consider $F(z) = R(f(x), h(x) - z)$ which is a polynomial in z, of degree at most n. It is desired to find the roots of $F(z)$ in $GF(q)$, i.e., to find those c's, $c \in GF(q)$ such that $F(c) = 0$, because for those c's it follows that $\gcd(f(x), h(x) - c) \neq 1$.

7.2.3 The Zassenhaus algorithm

Suppose that $c \in GF(q)$ and $c \in C$, where C contains all the c's such that $\gcd(f(x), h(x) - c) \neq 1$. From Theorem 6.7 it is known that

$$f(x) = \prod_{c \in C} \gcd(f(x), h(x) - c) \tag{7.1}$$

and also that

$$f(x) \left| \prod_{c \in C} (h(x) - c) \right. .$$

Let $G(z) = \prod_{c \in C}(z - c)$ and notice that $f(x)|G(h(x))$. The polynomial $G(z)$ is characterized in the following theorem.

THEOREM 7.2 *The polynomial $G(z)$ is the unique monic polynomial of least degree, among all the polynomials $g(z) \in GF(q)[z]$, such that $f(x)$ divides $g(h(x))$.*

Proof: The set of polynomials $g(z) \in \mathrm{GF}(q)[z]$, such that $f(x)$ divides $g(h(x))$, constitutes a nonzero ideal J of $\mathrm{GF}(q)[z]$. It follows that J is a principal ideal generated by $g_0(z) \in \mathrm{GF}(q)[z]$ which is a monic and uniquely determined polynomial. Consequently $G(z)$ is a multiple of $g_0(z)$ and thus $g_0(z) = \prod_{c \in C_1}(z - c)$, where $C_1 \subseteq C$, and $f(x)$ divides $g_0(h(x)) = \prod_{c \in C_1}(h(x) - c)$. Therefore, it follows that $f(x) = \prod_{c \in C_1} \gcd(f(x), h(x) - c)$ which in view of (7.1) leads to the conclusion that $C_1 = C$ and thus $g_0(z) = G(z)$.   □

Suppose the set C has size m and consider $G(z)$ expressed as

$$G(z) = \prod_{c \in C}(z - c) = \sum_{i=0}^{m} b_i z^i,$$

where $b_i \in \mathrm{GF}(q)$ and assuming $b_m = 1$. Since $f(x) | G(h(x))$ it follows that

$$\sum_{i=0}^{m} b_i [h(x)]^i = G(h(x)) \equiv 0 \mod f(x). \tag{7.2}$$

Expression (7.2) can be seen as a linear dependence relation among the residues of $1, h(x), h^2(x), \ldots, h^m(x)$ modulo $f(x)$. The polynomial $G(z)$ is determined by computing the residues modulo $f(x)$ of $[h(x)]^i$, $i = 0, 1, \ldots$, until a smallest power of $h(x)$ is found that is linearly dependent on the preceding powers of $h(x)$. The value of m is upperbounded by $m \leq k$, where k is the number of distinct prime polynomials in $f(x)$. The coefficients b_i, after normalization with $b_m = 1$, are the coefficients of $G(z) = \prod_{c \in C}(z - c)$. The elements $c \in C$ are the roots of $G(z)$.

This manner of finding the roots of $G(z)$ in $\mathrm{GF}(q)$ is known as the Zassenhaus algorithm.

7.3   Finding roots of polynomials over finite fields

It is now considered the problem of finding the roots in $\mathrm{GF}(q^m)$ of a polynomial of positive degree $f(x) \in \mathrm{GF}(q^m)[x]$. Polynomials over $\mathrm{GF}(q)$ can be viewed as polynomials over $\mathrm{GF}(q^m)$. By calculating $\gcd(f(x), x^r - x)$ it is obtained the part of $f(x)$ which factors as the product of monic linear polynomials over $\mathrm{GF}(r)$. Suppose that r is a prime number p. It is sufficient to consider only polynomials of the form

$$f(x) = \prod_{i=1}^{n}(x - c_i), \tag{7.3}$$

where the c_i's, $1 \leq i \leq n$, are distinct elements of $\mathrm{GF}(p)$. If p is small the roots of $f(x)$ can be found by exhaustive search through the c_i's. The case in which p is large is considered next.

7.3.1 Finding roots when p is large

Let $b \in \mathrm{GF}(p)$ and write (7.3) as

$$f(x - b) = \prod_{i=1}^{n} [x - (b + c_i)].$$

Notice that

$$f(x - b) | (x^p - x)$$

and that $x^p - x = x(x^{(p-1)/2} - 1)(x^{(p-1)/2} + 1)$. If $f(-b) = 0$ then x is a factor of $f(x - b)$ and a root of $f(x)$ has been found. Otherwise, write $f(x - b)$ as

$$f(x - b) = \gcd\left(f(x - b), x^{(p-1)/2} - 1\right) \gcd\left(f(x - b), x^{(p-1)/2} + 1\right)$$
(7.4)

and calculate

$$x^{(p-1)/2} \quad \mathrm{mod} \ f(x - b).$$

If $x^{(p-1)/2} \neq \pm 1$ modulo $f(x - b)$ then (7.4) provides a nontrivial factorization of $f(x)$. If $x^{(p-1)/2} \equiv \pm 1$ modulo $f(x - b)$ then a different value of b is selected and the procedure is repeated. Proceeding in this manner eventually all roots of $f(x)$ will be found. Notice that this procedure is not deterministic because the values for b are randomly chosen.

7.3.2 Finding roots when $q = p^m$ is large but p is small

Consider $f(x) = \Pi_{i=1}^{n}(x - \gamma_i)$, where $\gamma_i \in \mathrm{GF}(q)$, $1 \leq i \leq n$. Let $q = p^m$ and define the polynomial $S(x) = \sum_{j=0}^{m-1} x^{p^j}$ where, for $\gamma \in \mathrm{GF}(q)$, $S(\gamma)$ is the trace of γ over $\mathrm{GF}(p)$. It follows that $S(x)$ has degree p^{m-1} and the equation $S(\gamma) = c$, with $\gamma \in \mathrm{GF}(q)$ and $c \in \mathrm{GF}(p)$, can be shown to have p^{m-1} solutions (Lidl and Niederreiter 2006, p.152), i.e., p^{m-1} values of γ satisfying $S(\gamma) = c$, with $\gamma \in \mathrm{GF}(q)$. This remark leads to the following result:

$$x^q - x = \Pi_{c \in \mathrm{GF}(p)} \left(S(x) - c\right).$$

Since $f(x) | (x^q - x)$ it follows that $\Pi_{c \in \mathrm{GF}(p)} \left(S(x) - c\right) \equiv 0$ modulo $f(x)$, and thus

$$f(x) = \Pi_{c \in \mathrm{GF}(p)} \gcd(f(x), S(x) - c).$$

For small p the computation of p gcd's is not a problem. If, however, the factorization of $f(x)$ turn out to be trivial, i.e., if $S(x) \equiv c$ modulo $f(x)$ for some $c \in \mathrm{GF}(p)$, the following procedure is adopted.

Let $\{1, \beta, \beta^2, \ldots, \beta^{m-1}\}$ be a basis of GF(q) over GF(p). For $j = 0, 1, \ldots, m - 1$ substitute $\beta^j x$ for x in $x^q - x = \Pi_{c \in \mathrm{GF}(p)} (S(x) - c)$, leading to

$$(\beta^j)^q x^q - \beta^j x = \Pi_{c \in \mathrm{GF}(p)} \left(S(\beta^j x) - c\right)$$

or

$$x^q - x = \beta^{-j} \Pi_{c \in \mathrm{GF}(p)} \left(S(\beta^j x) - c\right).$$

The original factorization of $f(x)$ is now generalized and expressed as

$$f(x) = \Pi_{c \in \mathrm{GF}(p)} \gcd\left(f(x), S(\beta^j x) - c\right), \ 0 \le j \le m - 1.$$

It can be shown by contradiction that at least one of the above partial factorizations is nontrivial.

7.4 Problems with solutions

(1) Find the canonical factorization of $f(x) = x^5 + x^4 + 1$ in GF$(2)[x]$.

Solution: Notice that

(a) $\gcd(f(x), f'(x)) = 1$, meaning $f(x)$ has no repeated factor.

(b) Since N is not known, it is necessary to compute x^{2^r} modulo $(x^5 + x^4 + 1)$ until the condition $x^{2^r} \equiv x$ modulo $(x^5 + x^4 + 1)$ is satisfied, by trying $r = 1, 2, \ldots$

The operation of squaring a polynomial over GF(2) and reducing modulo $f(x)$ can be performed as a matrix multiplication as follows. Let $h(x) = h_0 + h_1 x + h_2 x^2 + h_3 x^3$, for example. Then it follows that

$$[h_0, h_1, h_2, h_3] \begin{bmatrix} x^0 & \mathrm{mod} \ f(x) \\ x^2 & \mathrm{mod} \ f(x) \\ x^4 & \mathrm{mod} \ f(x) \\ x^6 & \mathrm{mod} \ f(x) \end{bmatrix} = h_0 + h_1 x^2 + h_2 x^4 + h_3 x^6 \quad \mathrm{mod} \ f(x)$$

$$= h^2(x) \quad \mathrm{mod} \ f(x).$$

For $f(x) = x^5 + x^4 + 1$ it follows that

$$\begin{aligned} x^0 &\equiv 10000 & \mathrm{mod} \ f(x) \\ x^2 &\equiv 00100 & \mathrm{mod} \ f(x) \\ x^4 &\equiv 00001 & \mathrm{mod} \ f(x) \\ x^6 &\equiv 11011 & \mathrm{mod} \ f(x) \\ x^8 &\equiv 11111 & \mathrm{mod} \ f(x). \end{aligned}$$

It is now simpler to obtain the various powers x^{2^r} modulo $f(x)$ by matrix multiplication with reduction modulo $f(x)$, i.e.,

$$
\begin{aligned}
x &\equiv 01000 & \mod f(x) \\
x^2 &\equiv 00100 & \mod f(x) \\
x^4 &\equiv 00001 & \mod f(x) \\
x^8 &\equiv 11111 & \mod f(x) \\
x^{16} &\equiv 10011 & \mod f(x) \\
x^{32} &\equiv 10110 & \mod f(x) \\
x^{64} &\equiv 01000 & \mod f(x).
\end{aligned}
$$

Thus $x^{2^6} \equiv x$ modulo $(x^5 + x^4 + 1)$ and $N = 6$.

$$
\begin{aligned}
T_1(x) &= \sum_{j=0}^{5} x^{2^j} = x + x^2 + x^4 + x^8 + x^{16} + x^{32} \\
&\equiv 1 + x^2 + x^3 + x^4 \mod x^5 + x^4 + 1.
\end{aligned}
$$

$T_1(x)$ is thus f-reducing because it is not congruent to a constant modulo $f(x)$.

$$
\begin{aligned}
\gcd(x^5 + x^4 + 1, \ x^4 + x^3 + x^2 + 1) &= x^3 + x + 1 \\
\gcd(x^5 + x^4 + 1, \ x^4 + x^3 + x^2) &= x^2 + x + 1.
\end{aligned}
$$

Both factors, $x^3 + x + 1$ and $x^2 + x + 1$, are known to be irreducible in GF$(2)[x]$. Also, $N = 6 = \mathrm{lcm}(3, 2)$ confirming the result.

(2) Given $f(x) = a_3 x^3 + a_2 x^2 + a_1 x + a_0$ and $g(x) = b_2 x^2 + b_1 x + b_0$ calculate the resultant $R = R(f(x), g(x))$.

Solution: The resultant is calculated as follows:

$$
R = \begin{vmatrix}
a_3 & a_2 & a_1 & a_0 & 0 \\
0 & a_3 & a_2 & a_1 & a_0 \\
b_2 & b_1 & b_0 & 0 & 0 \\
0 & b_2 & b_1 & b_0 & 0 \\
0 & 0 & b_2 & b_1 & b_0
\end{vmatrix}
\quad
\begin{array}{l}
\left.\begin{array}{l} \cdot \\ \cdot \end{array}\right\} \text{2 rows} \\
\cdot \\
\left.\begin{array}{l} \cdot \\ \cdot \end{array}\right\} \text{3 rows}
\end{array}
$$

(3) Factor $f(x) = x^3 - x^2 + 1$ over GF(5).

Solution: By applying the Berlekamp algorithm it follows that $k = 2$ and that $h(x) = x^2 + x$. Then

$$F(z) = R(f(x), h(x) - z) = \begin{vmatrix} 1 & 4 & 0 & 1 & 0 \\ 0 & 1 & 4 & 0 & 1 \\ 1 & 1 & -z & 0 & 0 \\ 0 & 1 & 1 & -z & 0 \\ 0 & 0 & 1 & 1 & -z \end{vmatrix}$$

which produces $F(z) = -z^3 - 3z^2 - 1$ with roots 1 and 3. Thus,

$$\gcd(f(x), h(x) - 1) = x + 3$$
$$\gcd(f(x), h(x) - 3) = x^2 + x + 2.$$

Finally, $f(x) = (x + 3)(x^2 + x + 2)$.

(4) Find all f-reducing polynomials $h(x)$ for $f(x) = x^5 + 2x^4 + 5x^3 + 6x^2 - 3x + 10$ over GF(23).

Solution: The polynomials $h(x) = h_0 + h_1 x + h_2 x^2 + h_3 x^3 + h_4 x^4$ must satisfy the condition $h^{23}(x) \equiv h(x) \mod f(x)$, i.e., $h_0 + h_1 x^{23} + h_2 x^{46} + h_3 x^{69} + h_4 x^{92} \equiv h_0 + h_1 x + h_2 x^2 + h_3 x^3 + h_4 x^4$. Therefore, we now compute the following:

$$\begin{aligned}
x^{23} &\equiv 2x^4 + 21x^3 + 13x^2 + 11x + 12 \quad \mod f(x) \\
x^{46} &\equiv 2x^4 + 21x^3 + 14x^2 + 10x + 12 \quad \mod f(x) \\
x^{69} &\equiv x^3 \quad \mod f(x) \\
x^{92} &\equiv 22x^4 + 2x^3 + 10x^2 + 13x + 11 \quad \mod f(x)
\end{aligned}$$

From $h^{23}(x) - h(x) \equiv 0, \mod f(x)$, we construct the **B** matrix to satisfy $[h_0, h_1, h_2, h_3, h_4]\mathbf{B}^\mathsf{T} \equiv 0 \mod f(x)$.

$$\mathbf{B} = \begin{bmatrix} 1 & 12 & 12 & 0 & 11 \\ 0 & 11 & 10 & 0 & 13 \\ 0 & 13 & 14 & 0 & 10 \\ 0 & 21 & 21 & 1 & 2 \\ 0 & 2 & 2 & 0 & 22 \end{bmatrix} - \begin{bmatrix} 1 & 0 & 0 & 0 & 0 \\ 0 & 1 & 0 & 0 & 0 \\ 0 & 0 & 1 & 0 & 0 \\ 0 & 0 & 0 & 1 & 0 \\ 0 & 0 & 0 & 0 & 1 \end{bmatrix}$$

i.e,

$$\mathbf{B} = \begin{bmatrix} 0 & 12 & 12 & 0 & 11 \\ 0 & 10 & 10 & 0 & 13 \\ 0 & 13 & 13 & 0 & 10 \\ 0 & 21 & 21 & 0 & 2 \\ 0 & 2 & 2 & 0 & 21 \end{bmatrix}.$$

All rows in **B** are linearly dependent and the coefficients of $h(x)$ must satisfy the condition $h_0.0 + h_1.1 + h_2.1 + h_3.0 + h_4.22 = 0$. It follows that h_0 and h_3 can be chosen arbitrarily, while the condition $h_1 + h_2 = h_4 \mod 23$ implies that two out of these three variables can be chosen arbitrarily, and determine the value of the third variable. The dimension of the subspace of the h's is thus four, which means the $f(x)$ has four irreducible factors. The f-reducing polynomials $h(x)$ have the general form

$$h(x) = h_0 + h_1 x + h_2 x^2 + h_3 x^3 + h_4 x^4 = a + bx + cx^2 + dx^3 + (b+c)x^4,$$

where a, b, c and d take values in the closed interval $[0, 22]$.

(a) $\gcd(f(x), f'(x)) = 1$, meaning that $f(x)$ has no repeated factor.

(b) The Berlekamp algorithm gives $k = 4$ and for $a = b = d = 0$ and $c = 1$ the polynomial $h(x) = x^4 + x^2 \in GF(23)[x]$ is one of the f-reducing polynomials.

(c) The Zassenhaus algorithm is next applied to select the elements $c \in GF(23)$ for which $\gcd(f(x), h(x) - c) \neq 1$. First we need to compute $h^i(x), 1 \leq i \leq k$.

$$
\begin{aligned}
h(x) &\equiv x^4 + x^2 \mod f(x) \\
h^2(x) &\equiv 7x^4 + 14x^3 + 11x^2 + 19x + 3 \mod f(x) \\
h^3(x) &\equiv x^4 + 18x^3 + 9x^2 + 15x + 3 \mod f(x) \\
h^4(x) &\equiv 6x^4 + 3x^3 + 9x^2 + 20x + 13 \mod f(x)
\end{aligned}
$$

The desired linear dependence relation is given by

$$h^4(x) + 5h^3(x) - 5h^2(x) + h(x) + 10 \equiv 0 \mod f(x),$$

and thus

$$G(z) = z^4 + 5z^3 - 5z^2 + z + 10.$$

The roots of $G(z)$ in GF(23) are $-1, -3, -4$ and 3, and can be found using the theory developed in this chapter.

(5) Factor $x^5 + 2x^4 + 5x^3 + 6x^2 - 3x + 10$ over GF(23).

Solution: We use here the $h(x)$ computed in Problem 4. To find the explicit factorization of $f(x)$ over GF(23)[x] the following gcd's are computed.

$$\gcd(f(x), h(x) + 1) \;=\; x^2 + 22x + 1$$
$$\gcd(f(x), h(x) + 3) \;=\; x + 2$$
$$\gcd(f(x), h(x) + 4) \;=\; x + 11$$
$$\gcd(f(x), h(x) - 3) \;=\; x + 13.$$

Finally, $f(x)$ factors over GF(23) as

$$f(x) = (x^2 + 22x + 1)(x + 2)(x + 11)(x + 13).$$

(6) Find the roots of $f(x) = x^5 - x^4 + 2x^3 + x^2 - x - 2$, $f(x) \in \text{GF}(5)[x]$, contained in GF(5).

Solution:

(a) The roots of $f(x)$ in GF(5) are the same as the roots of $\gcd(x^5 - x, f(x)) = x^2 + x + 3$.

(b) Let $g(x) = x^2 + x + 3$. Calculating the residue of $x^{(5-1)/2} = x^2$ modulo $g(x)$ the result

$$x^2 \equiv 4x + 2 \quad \text{mod } g(x)$$

is obtained. This is equivalent to choosing $b = 0$ in the root finding algorithm. Also, since $x^{(5-1)/2} = x^2 \neq \pm 1$ modulo $g(x)$ the following nontrivial factorization of $g(x)$ is obtained.

$$\gcd(x^2 + x + 3, x^2 + 1) \;=\; x - 3$$
$$\gcd(x^2 + x + 3, x^2 - 1) \;=\; x - 1.$$

Thus, $g(x) = (x - 1)(x - 3)$ and the desired roots are 1 and 3.

(7) Prove that all the roots of $f(x) = x^3 + 8x^2 + 6x - 7 \in \text{GF}(19)[x]$ are contained in GF(19) and find them.

Solution:

(a) $\gcd(x^{19} - x, f(x)) = f(x) = x^3 + 8x^2 + 6x - 7.$

(b) $x^{(19-1)/2} = x^9 \equiv 10x^2 + 6x + 7$ modulo $f(x)$.

Proceeding as in the previous problem we obtain

$$\gcd(f(x), x^9 + 1) = x^2 + 12x - 3.$$

As a partial factorization of $f(x)$ we have $f(x) = (x-4)(x^2+12x-3)$, and also

$$\gcd(x^2 + 12x - 3, x^9 - 1) = x + 5,$$

with $x^2 + 12x - 3 = (x+5)(x+7)$. So, finally, $f(x) = (x-4)(x+5)(x+7)$, and the roots are $4, -5$ and -7.

Chapter 8

LINEARIZED POLYNOMIALS

8.1   Introduction

Linearized polynomials are useful in the study of polynomial factorization techniques because they have enough structure to allow an easy procedure to find their roots.

DEFINITION 8.1 *Over* $\mathrm{GF}(q^m)$, *a linearized polynomial (or a q-polynomial) is defined as a polynomial of the form*

$$L(x) = \sum_{i=0}^{n} \alpha_i x^{q^i},$$

with coefficients in the extension field $\mathrm{GF}(q^m)$ *of* $\mathrm{GF}(q)$.

8.2   Properties of $L(x)$

Let F be an arbitrary extension of $\mathrm{GF}(q^m)$.

THEOREM 8.2 *Let* $L(x) = \sum_{i=0}^{n} \alpha_i x^{q^i}$, $\alpha_i \in \mathrm{GF}(q^m)$, *then it follows that*

(a) Property 1:

$$L(\beta + \gamma) = L(\beta) + L(\gamma), \text{ for all } \beta, \gamma \in F.$$

(b) Property 2:

$$L(c\beta) = cL(\beta), \text{ for all } c \in \mathrm{GF}(q), \ \beta \in F.$$

Proof:

- Property 1.
  Given that $L(\beta) = \sum_{i=0}^{n} \alpha_i \beta^{q^i}$ and that $L(\gamma) = \sum_{i=0}^{n} \alpha_i \gamma^{q^i}$ it follows that

$$L(\beta + \gamma) = \sum_{i=0}^{n} \alpha_i (\beta + \gamma)^{q^i} = \sum_{i=0}^{n} \alpha_i (\beta^{q^i} + \gamma^{q^i}) = L(\beta) + L(\gamma).$$

- Property 2.
  Given that $L(c\beta) = \sum_{i=0}^{n} \alpha_i (c\beta)^{q^i}$ it follows that

$$
\begin{aligned}
L(c\beta) &= \sum_{i=0}^{n} \alpha_i (c\beta)^{q^i} = \sum_{i=0}^{n} \alpha_i c^{q^i} \beta^{q^i} = \sum_{i=0}^{n} \alpha_i c \beta^{q^i} \\
&= c \sum_{i=0}^{n} \alpha_i \beta^{q^i} = cL(\beta).
\end{aligned}
$$

$\square$

The following corollary is a consequence of properties 1 and 2 of $L(x)$.

COROLLARY 8.3 *For* $\beta_3 = c_1 \beta_1 + c_2 \beta_2$, $\beta_1 \in F$, $\beta_2 \in F$ *and* $c_1 \in$ GF(q), $c_2 \in$ GF(q) *it follows that*

$$L(\beta_3) = L(c_1 \beta_1 + c_2 \beta_2) = c_1 L(\beta_1) + c_2 L(\beta_2).$$

Because of properties 1 and 2 of $L(z)$, if F is considered as a vector space over GF(q), then $L(x)$ induces a linear operator on F.

8.3   Properties of the roots of $L(x)$

THEOREM 8.4 *Let* $L(x)$ *be a nonzero q-polynomial over* GF(q^m) *and let the extension field* GF(q^s) *of* GF(q^m) *contain all roots of* $L(x)$. *Then each root of* $L(x)$ *has the same multiplicity, which is either 1 or a power of q. Also, the roots form a linear subspace of* GF(q^s), *where* GF(q^s) *is regarded as a vector space over* GF(q).

Proof: From properties 1 and 2 it follows that any linear combination of roots with coefficients in GF(q) is again a root, and so the roots of $L(x)$ form a linear subspace of GF(q^s). If $L(x) = \sum_{i=0}^{n} \alpha_i x^{q^i}$, then $L'(x) = \alpha_0$. Thus, if $\alpha_0 \neq 0$ then $L(x)$ and $L'(x)$ are relatively prime and therefore $L(x)$ has simple roots only. Otherwise, assume

$\alpha_0 = \alpha_1 = \cdots = \alpha_{k-1} = 0$ and $\alpha_k \neq 0$, for some $k \geq 1$, and then

$$
\begin{aligned}
L(x) &= \sum_{i=k}^{n} \alpha_i x^{q^i} = \sum_{i=k}^{n} \alpha_i^{q^{mk}} x^{q^i} = \left[\sum_{i=k}^{n} \alpha_i^{q^{(m-1)k}} x^{q^{i-k}} \right]^{q^k} \\
&= \left[\sum_{j=0}^{n-k} \alpha_{j+k}^{q^{(m-1)k}} x^{q^j} \right]^{q^k} = \left[\sum_{j=0}^{n-k} \alpha_{j+k} x^{q^j} \right]^{q^k},
\end{aligned}
$$

i.e., $L(x)$ is expressed as the q^k-power of a linearized polynomial with simple roots only. Therefore, in this case, each root of $L(x)$ has multiplicity q^k. □

In order to proceed the following result is required.

LEMMA 8.5 *Let* $\beta_1, \beta_2, \ldots, \beta_n$ *be elements of* GF(q^m). *Then*

$$
|D_n| = \begin{vmatrix} \beta_1 & \beta_1^q & \cdots & \beta_1^{q^{n-1}} \\ \beta_2 & \beta_2^q & \cdots & \beta_2^{q^{n-1}} \\ \vdots & \vdots & & \vdots \\ \beta_n & \beta_n^q & \cdots & \beta_n^{q^{n-1}} \end{vmatrix} = \beta_1 \prod_{j=1}^{n-1} \prod_{c_j \in F_q} \left(\beta_{j+1} - \sum_{k=1}^{j} c_k \beta_k \right),
$$

and so, $|D_n| \neq 0$ *if and only if the* β_i, $1 \leq i \leq n$, *are linearly independent over* GF(q).

Proof: By induction on n the proof goes as follows. For $n = 1$, we have $|D_1| = \beta_1$ if we consider the empty product on the right-hand side equal to 1. Suppose the formula is true for some $n \geq 1$ and consider the polynomial

$$
D(x) = \begin{vmatrix} & & & \beta_1^{q^n} \\ & D_n & & \beta_2^{q^n} \\ & & & \vdots \\ x & x^q & \cdots & x^{q^n} \end{vmatrix} = x^{q^n} |D_n| + \sum_{i=0}^{n-1} \alpha_i x^{q^i}
$$

with $\alpha_i \in$ GF(q^m), $0 \leq i \leq n - 1$. Assume first that the $\beta_i, 1 \leq i \leq n$, are linearly independent over GF(q). We have $D(\beta^k) = 0, 1 \leq k \leq n$, because of repeated rows (i.e., linearly dependent rows) in the determinant. Also, since $D(x)$ is a q-polynomial over GF(q^m), all linear combinations of the roots, i.e.,

$$
\sum_{i=1}^{n} c_i \beta_i, \quad c_i \in \text{GF}(q), \ 1 \leq i \leq n,
$$

are roots of $D(x)$. Thus $D(x)$ has q^n distinct roots, and we obtain the factorization

$$D(x) = |D_n| \prod_{c_i \in F_q} \left(x - \sum_{i=1}^{n} c_i \beta_i \right). \qquad (8.1)$$

Now, if the β_i, $1 \le i \le n$, are linearly dependent over $\mathrm{GF}(q)$, then $|D_n| = 0$ and $\sum_{i=1}^{n} b_i \beta_i = 0$ for some choice of the $b_i \in \mathrm{GF}(q)$, $1 \le i \le n$, not all of which are zero. It then follows that

$$\sum_{i=1}^{n} b_i \beta_i^{q^j} = \left(\sum_{i=1}^{n} b_i \beta_i \right)^{q^j} = 0, \text{ for } j = 0, 1, \ldots, n,$$

and so the first n rows of $D(x)$ are also linearly dependent over $\mathrm{GF}(q)$. Thus $D(x) = 0$ and identity (8.1) is satisfied in all cases. Therefore,

$$|D_{n+1}| = D(\beta_{n+1}) = |D_n| \prod_{c_i \in F_q} \left(\beta_{n+1} - \sum_{i=1}^{n} c_i \beta_i \right),$$

which proves the lemma.   □

THEOREM 8.6 *Let U be a linear subspace of $\mathrm{GF}(q^m)$, considered as a vector space over $\mathrm{GF}(q)$. Then for any non-negative integer k, the polynomial*

$$L(x) = \prod_{\beta \in U} (x - \beta)^{q^k}$$

is a q-polynomial over $\mathrm{GF}(q^m)$.

Proof: Let $L_1(x) = \prod_{\beta \in U} (x - \beta)$. Let β_i, $1 \le i \le n$, be a basis for U over $\mathrm{GF}(q)$. By forming a determinant $|D_n|$ with these β_i, $1 \le i \le n$, as in Lemma 8.5, it follows from Lemma 8.5 that $|D_n| \ne 0$, and also that

$$L_1(x) = \prod_{\beta \in U} (x - \beta) = \prod_{c_i \in F_q} \left(x - \sum_{i=1}^{n} c_i \beta_i \right) = |D_n|^{-1} D(x)$$

from (8.1), and thus $L_1(x)$ is a q-polynomial over $\mathrm{GF}(q^m)$. Notice that $L(x) = L_1^{q^k}(x)$ is also a q-polynomial over $\mathrm{GF}(q^m)$ and the theorem is thus proved.   □

8.4 Finding roots of $L(x)$

Suppose that we are given $L(x) = \sum_{i=0}^{n} \alpha_i x^{q^i}$, a q-polynomial over $GF(q^m)$, and we want to find all the roots of $L(x)$ in the finite extension $GF(q^s)$ of $GF(q^m)$. As mentioned before, $L(x)$ induces a linear operator on $GF(q^s)$, i.e., the mapping $L : \beta \in GF(q^s) \to L(\beta) \in GF(q^s)$ is a linear operator on the vector space $GF(q^s)$ over $GF(q)$. Let $\{\beta_1, \beta_2, \ldots, \beta_s\}$ be a basis of $GF(q^s)$ over $GF(q)$, so that every $\beta \in GF(q^s)$ can be written as

$$\beta = \sum_{i=1}^{s} c_i \beta_i, \ c_i \in GF(q), \ 1 \le i \le s.$$

Then, by properties 1 and 2, we have

$$L(\beta) = L\left(\sum_{i=1}^{s} c_i \beta_i\right) = \sum_{i=1}^{s} c_i L(\beta_i).$$

Let $L(\beta_i) = \sum_{j=1}^{s} b_{ij} \beta_j$, $1 \le i \le s$, $b_{ij} \in GF(q)$, $1 \le i, j \le s$, and let $\mathbf{B}$ be the $s \times s$ matrix over $GF(q)$ whose (i, j) entry is b_{ij}. Then we can write

$$L(\beta) = \sum_{i=1}^{s} c_i L(\beta_i) = \sum_{i=1}^{s} c_i \sum_{j=1}^{s} b_{ij} \beta_j = \sum_{j=1}^{s} \sum_{i=1}^{s} c_i b_{ij} \beta_j = \sum_{j=1}^{s} d_j \beta_j,$$

if $d_j = \sum_{i=1}^{s} c_i b_{ij}$, or, in matrix terms as

$$(c_1, c_2, \ldots, c_s)\mathbf{B} = (d_1, d_2, \ldots, d_s),$$

where $\mathbf{B}$ is defined as

$$\mathbf{B} = \begin{bmatrix} b_{11} & b_{12} & \cdots & b_{1s} \\ b_{21} & b_{22} & \cdots & b_{2s} \\ \vdots & \vdots & & \vdots \\ b_{s1} & b_{s2} & \cdots & b_{ss} \end{bmatrix},$$

Therefore, the equation $L(\beta) = 0$ is equivalent to

$$(c_1, c_2, \ldots, c_s)\mathbf{B} = (0, 0, \ldots, 0).$$

This is a homogeneous system of s linear equations for the unknowns $(c_1, c_2, \ldots, c_s)$. If the rank of $\mathbf{B}$ is r then the system has q^{s-r} solution vectors $(c_1, c_2, \ldots, c_s)$. Each solution vector yields a root

$$\beta = \sum_{i=1}^{s} c_i \beta_i$$

of $L(x)$ in $GF(q^s)$. The problem of finding roots of $L(x)$ is thus converted into the problem of solving a homogeneous system of linear equations.

8.5 Affine *q*-polynomials

The method described earlier for finding roots of linearized polynomials can be extended to a more general class of polynomials called affine polynomials.

DEFINITION 8.7 *An affine q-polynomial over* GF(q^m) *is a polynomial of the form* $A(x) = L(x) - \alpha$, *where* $L(x)$ *is a linearized polynomial over* GF(q^m) *and* $\alpha \in$ GF(q^m).

It follows from Definition 8.7 that an element $\beta \in F$, where F is an extension of GF(q^m), is a root of $A(x)$ if and only if $L(\beta) = \alpha$, which by the notation employed in Section 8.4 is equivalent to

$$(c_1, c_2, \ldots, c_s)\mathbf{B} = (d_1, d_2, \ldots, d_s) \tag{8.2}$$

with $\alpha = \sum_{k=1}^{s} d_k \beta_k$. When the system of linear equations in (8.2) is solved for $(c_1, c_2, \ldots, c_s)$, each solution vector $(c_1, c_2, \ldots, c_s)$ yields a root $\beta = \sum_{j=1}^{s} c_j \beta_j$ of $A(x)$ in F. Finding roots of affine polynomials is therefore an easier task than finding roots of polynomials $f(x)$ over GF(q^m) in general.

It is described next a method for finding roots of an arbitrary polynomial over a finite field F, which is an extension of GF(q^m), using previous knowledge about affine polynomials as follows.

(a) Determine a nonzero affine *q*-polynomial $A(x)$ over GF(q^m) that contains $f(x)$ as one of its factors. The polynomial $A(x)$ is called the affine multiple of $f(x)$.

(b) Determine all the roots of $A(x)$ in F, as described earlier.

(c) Calculate $f(\beta)$ for all values of β such that $\beta \in F$ and $A(\beta) = 0$, selecting those β such that $f(\beta) = 0$. This step follows from the observation that $f(x)$ is a factor of $A(x)$ and thus the roots of $f(x)$ in F are contained in the set of roots of $A(x)$ in F.

In order to implement this procedure, the following question needs to be answered. How can $A(x)$ be found such that $A(x) = f(x)g(x)$? One possible answer is as follows.

Let $f(x)$ be a polynomial of degree $n \geq 1$.

(a) For $i = 0, 1, \ldots, n-1$ calculate the unique polynomial $r_i(x)$ of degree at most $n - 1$ such that

$$x^{q^i} \equiv r_i(x) \bmod f(x).$$

(b) Choose $\alpha_i \in \mathrm{GF}(q^m)$, $0 \le i \le n-1$, where $\alpha_i \ne 0$ for at least one value of i, such that $\sum_{i=0}^{n-1} \alpha_i r_i(x)$ is a constant polynomial.

(c) Once a nontrivial solution has been found, i.e., $\sum_{i=0}^{n-1} \alpha_i r_i(x) = \alpha$ for some $\alpha \in \mathrm{GF}(q^m)$, it follows that:

$$\sum_{i=0}^{n-1} \alpha_i x^{q^i} \equiv \sum_{i=0}^{n-1} \alpha_i r_i(x) = \alpha \bmod f(x)$$

and thus

$$A(x) = \sum_{i=0}^{n-1} \alpha_i x^{q^i} - \alpha$$

is a nonzero affine q-polynomial over $\mathrm{GF}(q^m)$ which is a multiple of $f(x)$.

8.6   Problems with solutions

(1) Consider the 2-polynomial $L(x) = x^{16} + x^8 + \alpha x^4$ over $\mathrm{GF}(4)$, where α is a root of the primitive polynomial $x^2 + x + 1$ over $\mathrm{GF}(2)$, i.e., $\alpha \in \mathrm{GF}(2^2) = \mathrm{GF}(4)$ and $\alpha^2 + \alpha + 1 = 0$. Consider $\mathrm{GF}(2^4)$ as an extension of $\mathrm{GF}(4)$. Which are the roots of $L(x)$ contained in $\mathrm{GF}(16)$?

Solution: Consider the basis $\{1, \beta, \beta^2, \beta^3\}$ of $\mathrm{GF}(2^4)$ over $\mathrm{GF}(2)$, where β is a root of the primitive polynomial $x^4 + x + 1$ over $\mathrm{GF}(2)$. The relation between α and β is the following. Since $\alpha^3 = 1$ and $\beta^{15} = 1$, the value $\alpha = \beta^5$ can be chosen (other possible value is $\alpha = \beta^{10}$), i.e., $\alpha = \beta^2 + \beta$ modulo $(\beta^4 + \beta + 1)$. It follows that

$$L(x) = x^{16} + x^8 + \alpha x^4 = (x^4 + x^2 + \alpha x)^4 = L_1^4(x),$$

where $L_1(x) = x^4 + x^2 + \alpha x$ and $\alpha = \beta^5 = \beta^2 + \beta$ modulo $\beta^4 + \beta + 1$ is considered, and

$$
\begin{aligned}
L_1(1) &= \alpha = \beta^2 + \beta \\
L_1(\beta) &= \beta^4 + \beta^2 + \alpha\beta = \beta^3 + \beta + 1 \\
L_1(\beta^2) &= \beta^8 + \beta^4 + \alpha\beta^2 = \beta^3 + \beta^2 + 1 \\
L_1(\beta^3) &= \beta^{12} + \beta^6 + \alpha\beta^3 = \beta^2 + \beta,
\end{aligned}
$$

$$
\mathbf{B} = \begin{bmatrix} 0 & 1 & 1 & 0 \\ 1 & 1 & 0 & 1 \\ 1 & 0 & 1 & 1 \\ 0 & 1 & 1 & 0 \end{bmatrix}.
$$

The system of linear equations $(c_1, c_2, c_3, c_4)\mathbf{B} = (0, 0, 0, 0)$ leads to the following set of equations:

$$
\begin{array}{rcl}
c_2 + c_3 && = 0 \\
c_1 + c_2 + c_4 && = 0 \\
c_1 + c_3 + c_4 && = 0
\end{array}
\quad\longrightarrow\quad
\begin{array}{rcl}
c_2 + c_3 &=& 0 \\
c_2 &=& c_1 + c_4 \\
c_3 &=& c_1 + c_4
\end{array}
$$

with solutions $(0, 0, 0, 0)$ and $(1, 0, 0, 1)$ for $c_2 = c_3 = 0$, and $(0, 1, 1, 1)$ and $(1, 1, 1, 0)$ for $c_2 = c_3 = 1$. The roots of $L_1(x)$ in GF(16), generally denoted as $\delta = \sum_{i=1}^{3} c_i \beta_i$, are the following:

$$
\begin{array}{rcl}
\delta_1 &=& 0 \\
\delta_2 &=& \beta^3 + 1 = \beta^{14} \\
\delta_3 &=& \beta^3 + \beta^2 + \beta = \beta^{11} \\
\delta_4 &=& \beta^2 + \beta + 1 = \beta^{10}.
\end{array}
$$

(2) Let $f(x) = x^4 + \alpha^2 x^3 + \alpha x^2 + \alpha^2 x + 1$, $f(x) \in$ GF(4)$[x]$, where α is a root of $x^2 + x + 1$ which is primitive over GF(2). Find the roots of $f(x)$ belonging to GF(16).

Solution:

(a) Finding $A(x)$, an affine multiple of $f(x)$, with $q = 2$.

$$
\begin{array}{rcl}
x &=& r_0(x) \\
x^2 &=& r_1(x) \\
x^4 &\equiv& \alpha^2 x^3 + \alpha x^2 + \alpha^2 x + 1 \bmod f(x) = r_2(x) \\
x^8 &\equiv& x^3 + \alpha x^2 + \alpha^2 x + 1 \bmod f(x) = r_3(x)
\end{array}
$$

Consider the polynomial

$$
\alpha_0 r_0(x) + \alpha_1 r_1(x) + \alpha_2 r_2(x) + \alpha_3 r_3(x) = \theta \qquad (8.3)
$$

i.e., a constant polynomial. Replacing $r_0(x), r_1(x), r_2(x)$ and $r_3(x)$ in (8.3) by their corresponding values it follows that

$$
\alpha_0 x + \alpha_1 x^2 + \alpha_2 (\alpha^2 x^3 + \alpha x^2 + \alpha^2 x + 1) + \alpha_3 (x^3 + \alpha x^2 + \alpha^2 x + 1) = \theta
$$

from which it follows:

$$
\begin{array}{rcl}
\alpha_1 + \alpha\alpha_2 + \alpha\alpha_3 &=& 0 \\
\alpha_0 + \alpha^2\alpha_2 + \alpha^2\alpha_3 &=& 0 \\
\alpha^2\alpha_2 + \alpha_3 &=& 0 \\
\alpha_2 + \alpha\alpha_3 &=& \theta.
\end{array}
$$

Choosing $\alpha_3 = 1$ the solution $(\alpha_0, \alpha_1, \alpha_2, \alpha_3) = (\alpha, 1, \alpha, 1)$ and $\theta = 0$ is obtained, and thus $A(x) = \alpha_3 x^8 + \alpha_2 x^4 + \alpha_1 x^2 + \alpha_0 x - \theta$ is written explicitly as

$$A(x) = x^8 + \alpha x^4 + x^2 + \alpha x.$$

(b) The roots of $A(x)$ in GF(16) are calculated next. Since $\theta = 0$, the problem reduces to finding the roots of $L(x) = 0$ (a polynomial in GF(4)) belonging to GF(16), where $L(x) = A(x)$. Use $\alpha = \beta^5$ where β is a primitive element of GF(16) satisfying the relation $\beta^4 = \beta + 1$. It follows that

$$\left. \begin{array}{rcl} L(1) & = & 0 \\ L(\beta) & = & \beta^2 + \beta + 1 \\ L(\beta^2) & = & \beta^2 + \beta + 1 \\ L(\beta^3) & = & \beta^2 + \beta + 1 \end{array} \right\} \longrightarrow (c_1, c_2, c_3, c_4)\mathbf{B} = (0,0,0,0)$$

with

$$\mathbf{B} = \begin{bmatrix} 0 & 0 & 0 & 0 \\ 1 & 1 & 1 & 0 \\ 1 & 1 & 1 & 0 \\ 1 & 1 & 1 & 0 \end{bmatrix}.$$

The matrix $\mathbf{B}$ has rank 1 and thus $L(x)$ has $2^3 = 8$ roots in GF(16), namely $\delta_1 = 0$, $\delta_2 = \beta + \beta^2 = \beta^5 = \alpha$, $\delta_3 = \beta + \beta^3 = \beta^9$, $\delta_4 = \beta^2 + \beta^3 = \beta^6$, $\delta_5 = 1$, $\delta_6 = \beta^{10} = \alpha^2$, $\delta_7 = \beta^7$, $\delta_8 = \beta^{13}$. The roots of $f(x)$ are δ_2, δ_3, δ_4 and δ_6, i.e., all roots of $f(x)$ lie in GF(16).

Chapter 9

GOPPA CODES

9.1   Introduction

The study of Goppa codes (Goppa 1970) is important for at least the following reasons.

(a) Goppa codes generalize the narrow sense BCH codes.

(b) Their class contains arbitrarily long q-ary codes the d_{min}/n of which strictly exceeds the asymptotic Gilbert bound for all $q \geq 49$ and for every rate R in a certain interval depending on q.

(c) They can be efficiently decoded up to their designed distance.

(d) They have been proposed for a public-key cryptosystem (McEliece 1978).

Before Goppa codes are formally defined some of the classic theory of BCH codes is reviewed, however, cast in a light appropriate to what is needed later.

LEMMA 9.1   *The n-tuple $\mathbf{c} = (c_0, c_1, \ldots, c_{n-1}) \in \mathrm{GF}(q)^n$ is a codeword of the narrow sense BCH code over $\mathrm{GF}(q)$ defined by the roots $\alpha, \alpha^2, \ldots, \alpha^{d-1}$ of the generator polynomial $g(x)$ if and only if*

$$\sum_{i=0}^{n-1} c_i \alpha^{i(d-1)} \left[\frac{x^{d-1} - \alpha^{-i(d-1)}}{x - \alpha^{-i}} \right] = 0. \tag{9.1}$$

Proof: From the definition of narrow sense BCH codes (MacWilliams and Sloane 1977, p.28) it follows that $\mathbf{c}$ is a codeword in this code if and only if

$$\sum_{i=0}^{n-1} c_i \alpha^{ij} = 0, \text{ for } 1 \le j \le d - 1.$$

However, by manipulating the left-hand side of (9.1) it follows that

$$\sum_{i=0}^{n-1} c_i \alpha^{i(d-1)} \left[\frac{x^{d-1} - \alpha^{-i(d-1)}}{x - \alpha^{-i}} \right] = \sum_{i=0}^{n-1} c_i \alpha^{i(d-1)} \sum_{j=0}^{d-2} \alpha^{-i(d-2-j)} x^j$$

$$= \sum_{j=0}^{d-2} \sum_{i=0}^{n-1} c_i \alpha^{i(j+1)} x^j$$

$$= \sum_{j=1}^{d-1} \left(\sum_{i=0}^{n-1} c_i \alpha^{ij} \right) x^{j-1} = 0.$$

Notice that for this last polynomial to be identically zero each of its coefficients has to be zero. Thus $\mathbf{c}$ is a codeword if and only if (9.1) holds and the lemma thus follows. $\square$

9.2 Parity-check equations

It is remarked that (9.1) defines a set of parity-check equations and from them the code parity-check matrix is derived as now explained. An intermediate result in the proof of Lemma 9.1 was the following:

$$\sum_{j=1}^{d-1} \left(\sum_{i=0}^{n-1} c_i \alpha^{ij} \right) x^{j-1} = 0,$$

where,

$$\sum_{i=0}^{n-1} c_i \alpha^{ij} = 0, \text{ for } 1 \le j \le d - 1. \tag{9.2}$$

In matrix form (9.2) can be written as

$$(c_0, c_1, \ldots, c_{n-1}) \begin{bmatrix} 1 & \alpha & \alpha^2 & \cdots & \alpha^{n-1} \\ 1 & \alpha^2 & \alpha^4 & \cdots & \alpha^{2(n-1)} \\ \cdot & \cdot & \cdot & \cdots & \cdot \\ \cdot & \cdot & \cdot & \cdots & \cdot \\ 1 & \alpha^{d-1} & \alpha^{2(d-1)} & \cdots & \alpha^{(d-1)(n-1)} \end{bmatrix}^{\mathrm{T}} = [\mathbf{0}],$$

where $[\mathbf{0}]$ denotes a $(d-1)$-place all-zero row matrix. Alternatively, this result can be expressed as $\mathbf{c}\mathbf{H}^{\mathrm{T}} = \mathbf{0}$, where $\mathbf{H} = [\alpha^{ij}]$, $0 \le i \le n-1$, $1 \le j \le d - 1$, and α is a primitive n^{th} root of unity.

DEFINITION 9.2 **Goppa codes:** *Let $g(x)$ be a polynomial of degree t, $1 \leq t < n$, with coefficients on an extension field $GF(q^m)$ of $GF(q)$, and let $L = \{\gamma_0, \gamma_1, \ldots, \gamma_{n-1}\}$ be a set of distinct elements of $GF(q^m)$ such that $g(\gamma_i) \neq 0$, $0 \leq i \leq n-1$. The Goppa code $\Gamma(L, g)$ over $GF(q)$, with Goppa polynomial $g(x)$, is the set of all n-tuples $\mathbf{c} = (c_0, c_1, \ldots, c_{n-1}) \in GF(q)^n$, i.e., with symbols from $GF(q)$, such that the following identity*

$$\sum_{i=0}^{n-1} c_i g^{-1}(\gamma_i) \left[\frac{g(x) - g(\gamma_i)}{x - \gamma_i} \right] = 0 \qquad (9.3)$$

holds in the polynomial ring $GF(q^m)[x]$. If $g(x)$ is irreducible over $GF(q^m)$, then $\Gamma(L, g)$ is called an irreducible Goppa code.

If (9.3) reduced modulo $g(x)$ is considered, an equivalent expression for the definition of Goppa codes will result, i.e., the code is defined by the set of all $GF(q)$ n-tuples that satisfy the congruence

$$\sum_{i=0}^{n-1} \frac{c_i}{x - \gamma_i} \equiv 0 \pmod{g(x)}. \qquad (9.4)$$

EXAMPLE 9.3 *Let $g(x) = x^{d-1}$ and let $L = \alpha^{-i}$, $0 \leq i \leq n-1$, where $\alpha \in GF(q^m)$ is a primitive n^{th} root of unity, then $\Gamma(L, g)$ is defined by*

$$\sum_{i=0}^{n-1} c_i \alpha^{i(d-1)} \left[\frac{x^{d-1} - \alpha^{-i(d-1)}}{x - \alpha^{-i}} \right]$$

which is exactly expression (9.1), i.e., the given $\Gamma(L, g)$ defines a narrow sense BCH code over $GF(q)$ of block length n (and designed distance d). The parenthesis was used to emphasize that knowledge about d was not derived from the definition of $\Gamma(L, g)$.

9.3 Parity-check matrix of Goppa codes

By expanding the left-hand side of (9.4) in powers of x, reduced modulo $g(x)$, and then equating the coefficients to zero, the result is a set of $\deg(g(x))$ linear equations in $GF(q^m)$, involving the codeword components c_i, which can be viewed as defining a set of parity-check equations. It is therefore concluded that Goppa codes are linear codes. Expression (9.4) can be seen to define a subspace of the vector space $GF(q)^n$, i.e., it restricts the $\mathbf{c} \in GF(q)^n$ to those satisfying (9.4).

It is now of interest to find a parity-check matrix $\mathbf{H}$ for $\Gamma(L, g)$, i.e., a matrix $\mathbf{H}$ over $GF(q^m)$, such that the intersection of its null space with $GF(q)^n$ is $\Gamma(L, g)$. Let $g(x) = \sum_{j=0}^{t} g_j x^j$, then it follows that

$$\frac{g(x) - g(\gamma)}{x - \gamma} = \sum_{j=0}^{t} g_j \frac{x^j - \gamma^j}{x - \gamma} = \sum_{j=0}^{t} g_j \sum_{s=0}^{j-1} \gamma^{(j-1-s)} x^s$$

$$= \sum_{s=0}^{t-1} \sum_{j=s+1}^{t} g_j \gamma^{(j-1-s)} x^s. \tag{9.5}$$

By making $h_i = g^{-1}(\gamma_i)$, $0 \leq i \leq n - 1$, and using (9.4), it follows that (9.3) is satisfied by $\mathbf{c} \in \mathrm{GF}(q)^n$ if and only if

$$\sum_{s=0}^{t-1} \sum_{i=0}^{n-1} c_i \left(h_i \sum_{j=s+1}^{t} g_j \gamma_i^{(j-1-s)} \right) x^s = 0,$$

i.e., if and only if

$$\sum_{i=0}^{n-1} c_i \left(h_i \sum_{j=s+1}^{t} g_j \gamma_i^{(j-1-s)} \right) = 0, \text{ for } 0 \leq s \leq t - 1. \tag{9.6}$$

In matrix form (9.6) can be expressed as

$$\mathbf{c} \begin{bmatrix} h_0 g_t & \cdots & h_{n-1} g_t \\ h_0(\gamma_0 g_t + g_{t-1}) & \cdots & h_{n-1}(\gamma_{n-1} g_t + g_{t-1}) \\ . & \cdots & . \\ . & \cdots & . \\ h_0 \sum_{j=1}^{t} g_j \gamma_0^{j-1} & \cdots & h_{n-1} \sum_{j=1}^{t} g_j \gamma_{n-1}^{j-1} \end{bmatrix}^{\mathrm{T}} = [\mathbf{0}],$$

i.e., $\mathbf{c}\mathbf{H}^{\mathrm{T}} = [\mathbf{0}]$, where for $\mathbf{H}$, since $g_t \neq 0$, after some elementary row operations it follows that

$$\mathbf{H} = \begin{bmatrix} h_0 & h_1 & \cdots & h_{n-1} \\ h_0\gamma_0 & h_1\gamma_1 & \cdots & h_{n-1}\gamma_{n-1} \\ . & . & \cdots & . \\ . & . & \cdots & . \\ h_0\gamma_0^{t-1} & h_1\gamma_1^{t-1} & \cdots & h_{n-1}\gamma_{n-1}^{t-1} \end{bmatrix}. \tag{9.7}$$

The intersection of the null space of $\mathbf{H}$ in (9.7) with $\mathrm{GF}(q)^n$, is $\Gamma(L, g)$. The elements of $\mathbf{H}$ belong to $\mathrm{GF}(q^m)$. By mapping each $\mathrm{GF}(q^m)$ entry in $\mathbf{H}$ as a column m-tuple over $\mathrm{GF}(q)$ an equivalent (over $\mathrm{GF}(q)$) parity-check matrix results.

THEOREM 9.4 *The dimension k of the Goppa code $\Gamma(L, g)$ is at least $n - mt$ and its minimum distance is at least $t + 1$.*

Proof: Over GF(q^m), **H** is a $t \times n$ matrix. When each entry of **H** is mapped as a column m-tuple over GF(q) a matrix with rank at most mt results. Thus, it follows that $k \geq n - mt$. For the minimum distance of $\Gamma(L, g)$ it is noticed that, in a way similar as done for BCH codes, any t columns of **H** can be chosen and, after simplification, a Vandermonde determinant can be formed with the powers of t distinct values of $\gamma_i, 0 \leq i \leq n - 1$. Therefore such a determinant is nonzero and any t columns of **H** are linearly independent. Thus the minimum distance of $\Gamma(L, g)$ satisfies $d \geq t + 1$. $\qquad\qquad\square$

In practice most of the codes used are binary. For binary Goppa codes the lower bound on the minimum distance can be improved as shown next.

THEOREM 9.5 *A binary Goppa code whose Goppa polynomial has degree t and contains no repeated irreducible factors over* GF(2^m), *has a minimum distance of at least $2t + 1$.*

Proof: Let $\mathbf{c} \in \text{GF}(2)^n$ be a codeword with Hamming weight w in the binary Goppa code $\Gamma(L, g)$ and let $c_{i_1} = c_{i_2} = \cdots = c_{i_w} = 1$, with $0 \leq i_1 \leq i_2 \leq \ldots \leq i_w \leq n - 1$, be the nonzero positions of **c**. Let $L = \{\gamma_0, \gamma_1, \ldots, \gamma_{n-1}\} \subseteq \text{GF}(2^m)$, and define

$$f(x) = \prod_{j=1}^{w}(x - \gamma_{i_j}) \in \text{GF}(2^m)[x].$$

By multiplying (9.4) by $f(x)$ it follows that

$$\sum_{j=1}^{w} \prod_{s=1, s \neq j}^{w}(x - \gamma_{i_s}) = f'(x) = 0 \mod g(x),$$

which means that $g(x)$ divides $f'(x)$. In fields of characteristic 2 the derivative of a polynomial $f(x)$, i.e., $f'(x)$, contains only even powers of x and is therefore the square of a polynomial in GF(2^m)[x]. Since, by hypothesis, $g(x)$ has no repeated irreducible factors over GF(2^m) it follows that $g^2(x)$ divides $f'(x)$, which implies that

$$w - 1 \geq \deg(f'(x)) \geq 2\deg(g(x)) = 2t,$$

i.e., $w - 1 \geq 2t$. Thus $w \geq 2t + 1$ and so the minimum weight of any non-zero codeword is at least $2t + 1$, which by the code linearity also implies $d_{\min} \geq 2t + 1$. $\qquad\qquad\square$

It is remarked that this may not be a very good lower bound on $d_{\min}$ but the $\deg(g(x))$ plays for Goppa codes a more significant role than the weight of the generator polynomial $g(x)$ of ordinary cyclic codes.

9.4   Algebraic decoding of Goppa codes

In general, the minimum distance of Goppa codes is not known exactly. Theorems 9.4 and 9.5 provide only lower bounds on the minimum distance. In the sequel procedures for decoding Goppa codes will be presented, conditioned on the knowledge of lower bounds on minimum distance.

The algebraic decoding of Goppa codes can be performed by using the Euclidean division algorithm for polynomials. When decoded in this manner, a Goppa code over $GF(q)$, with generator polynomial of degree $2t$, corrects up to t errors. In particular, over $GF(2)$, a Goppa code with generator polynomial of degree t can be decoded with the Euclidean algorithm to correct up to t errors. Goppa codes with generator polynomial $g(x) = x^{2t}$ can be efficiently decoded with the Berlekamp–Massey algorithm (Massey 1969). Let $\mathbf{c}$ be the transmitted codeword, let $\mathbf{e}$ be the error vector added by the channel and let $\mathbf{r} = \mathbf{c} + \mathbf{e}$ be the received n-tuple. It follows that

$$\sum_{i=0}^{n-1} \frac{r_i}{x - \gamma_i} = \sum_{i=0}^{n-1} \frac{c_i}{x - \gamma_i} + \sum_{i=0}^{n-1} \frac{e_i}{x - \gamma_i}.$$

However, since $\mathbf{c}$ is a codeword it follows from (9.4) that $\sum_{i=0}^{n-1} \frac{c_i}{x-\gamma_i} \equiv 0$ mod $g(x)$ and it also follows that

$$\sum_{i=0}^{n-1} \frac{r_i}{x - \gamma_i} \equiv \sum_{i=0}^{n-1} \frac{e_i}{x - \gamma_i} \quad \text{mod } g(x). \tag{9.8}$$

The syndrome polynomial $S(x)$ is defined as

$$S(x) \equiv \sum_{i=0}^{n-1} \frac{r_i}{x - \gamma_i} \quad \text{mod } g(x), \tag{9.9}$$

where it is noticed that the degree of $S(x)$ is less than $\deg(g(x))$. By combining (9.8) and (9.9) it follows that

$$S(x) \equiv \sum_{i=0}^{n-1} \frac{e_i}{x - \gamma_i} \quad \text{mod } g(x).$$

Let M be the subset of L such that $e_i \neq 0$ if and only if $\gamma_i \in M$, then

$$S(x) \equiv \sum_{\gamma_i \in M} \frac{e_i}{x - \gamma_i} \quad \text{mod } g(x). \tag{9.10}$$

The error locator polynomial $\sigma(x)$ is defined as

$$\sigma(x) = \prod_{\gamma_i \in M} (x - \gamma_i). \tag{9.11}$$

The error evaluator polynomial $\eta(x)$ is defined as

$$\eta(x) = \sum_{\gamma_i \in M} e_i \prod_{\gamma_j \in M, j \neq i} (x - \gamma_j). \tag{9.12}$$

It is remarked that $\sigma(x)$ and $\eta(x)$ are relatively prime polynomials. The formal derivative of (9.11) gives

$$\sigma'(x) = \sum_{\gamma_i \in M} \prod_{\gamma_j \in M, j \neq i} (x - \gamma_j). \tag{9.13}$$

Hence, for each $\gamma_i \in M$ it follows that

$$\eta(\gamma_i) \equiv e_i \prod_{\gamma_j \in M, j \neq i} (\gamma_i - \gamma_j) = e_i \sigma'(\gamma_i),$$

and thus, $e_i = \eta(\gamma_i)/\sigma'(\gamma_i)$.

The major effort in the algebraic decoding operation is spent to obtain the locator polynomial $\sigma(x)$ and the evaluator polynomial $\eta(x)$. After that is done the error vector is fully determined from

$$e_i = \begin{cases} 0 & , \quad \text{if } \sigma(\gamma_i) \neq 0 \\ \eta(\gamma_i)/\sigma'(\gamma_i) & , \quad \text{if } \sigma(\gamma_i) = 0. \end{cases}$$

By multiplying (9.10) and (9.11) the right-hand side of (9.12) is obtained, i.e.,

$$S(x)\sigma(x) \equiv \eta(x) \mod g(x), \tag{9.14}$$

which is called the *key equation* for decoding Goppa codes (Berlekamp 1973).

Given $g(x)$ and $S(x)$, the task of an algebraic decoding algorithm is to find low-degree polynomials $\sigma(x)$ and $\eta(x)$ satisfying (9.14). Expression (9.14) can be written as a system of $\deg(g(x))$ linear equations in the unknown coefficients of $\sigma(x)$ and $\eta(x)$, by reducing each power of x mod $g(x)$ and equating coefficients of the same degree on both sides.

In order to prove that the decoder can correct up to t errors, it is sufficient to show that the system of linear equations in (9.14) has at most one solution with $\deg(\sigma(x))$ and $\deg(\eta(x))$ sufficiently small.

Suppose that $S(x)\sigma(x) \equiv \eta(x) \mod g(x)$ has two distinct pairs of solutions denoted by

$$S(x)\sigma^{(1)}(x) \equiv \eta^{(1)}(x) \mod g(x), \tag{9.15}$$
$$S(x)\sigma^{(2)}(x) \equiv \eta^{(2)}(x) \mod g(x), \tag{9.16}$$

where $\gcd[\sigma^{(1)}(x), \eta^{(1)}(x)] = 1$ and $\gcd[\sigma^{(2)}(x), \eta^{(2)}(x)] = 1$. If $\sigma^{(1)}(x)$ and $g(x)$ had a common factor then $\eta^{(1)}(x)$ would be divisible by that factor contradicting the fact that $\sigma^{(1)}(x)$ and $\eta^{(1)}(x)$ are relatively prime. Thus, dividing (9.15) by $\sigma^{(1)}(x)$ it follows that

$$S(x) \equiv \frac{\eta^{(1)}(x)}{\sigma^{(1)}(x)} \quad \mathrm{mod}\ g(x),$$

and proceeding similarly with (9.16) it follows that

$$S(x) \equiv \frac{\eta^{(2)}(x)}{\sigma^{(2)}(x)} \quad \mathrm{mod}\ g(x),$$

and therefore

$$\sigma^{(1)}(x)\eta^{(2)}(x) \equiv \sigma^{(2)}(x)\eta^{(1)}(x) \quad \mathrm{mod}\ g(x). \tag{9.17}$$

If $\deg(g(x)) = 2t$ and $\deg(\sigma^{(1)}(x)) \le t$, $\deg(\sigma^{(2)}(x)) \le t$, $\deg(\eta^{(1)}(x)) < t$, $\deg(\eta^{(2)}(x)) < t$, then an equality in (9.17) follows, i.e.,

$$\sigma^{(1)}(x)\eta^{(2)}(x) = \sigma^{(2)}(x)\eta^{(1)}(x).$$

Since $\gcd[\sigma^{(1)}(x),\ \eta^{(1)}(x)] = 1$ then $\sigma^{(1)}(x)$ divides $\sigma^{(2)}(x)$ and similarly, if $\gcd[\sigma^{(2)}(x), \eta^{(2)}(x)] = 1$ then $\sigma^{(2)}(x)$ divides $\sigma^{(1)}(x)$. The σ's are monic polynomials and thus it follows that $\sigma^{(1)}(x) = \sigma^{(2)}(x)$ and $\eta^{(1)}(x) = \eta^{(2)}(x)$. Our conclusion is that if $\deg(g(x)) = 2t$ then $S(x)\sigma(x) \equiv \eta(x) \quad \mathrm{mod}\ g(x)$ has at most one solution with $\deg(\eta(x)) < \deg(\sigma(x)) \le t$. This also implies that the associated system of linear equations in the unknown coefficients of $\sigma(x)$ and $\eta(x)$ must be nonsingular.

9.4.1   The Patterson algorithm

In this section, we study the Patterson algorithm (Patterson 1975) for decoding binary Goppa codes when the generator polynomial $g(x)$, of degree t, is irreducible over $\mathrm{GF}(2^m)$. It is known (MacWilliams and Sloane 1977, p.366) that the congruence

$$S(x)\sigma(x) = \eta(x) \quad \mathrm{mod}\ g(x),$$

has at most one solution with $\deg(\sigma(x)) \le t$ and $\deg(\eta(x)) < t$, if $\deg(g(x)) = 2t$. For binary irreducible Goppa codes, with $\deg(g(x)) = t$, it is known that $d \ge 2t + 1$ (p.345).

The following procedure, as described in (Patterson 1975), allows the correction of up to t errors. Let

$$\sigma(x) = \alpha^2(x) + x\beta^2(x), \tag{9.18}$$

and thus $\sigma'(x) = \beta^2(x)$, and $\deg(\alpha(x)) \le t/2$ and $\deg(\beta(x)) \le (t-1)/2$. Because $g(x)$ is irreducible, by the Euclidean division algorithm for polynomials $h(x)$ can be found, with $\deg(h(x)) < t$, such that

$$S(x)h(x) \equiv 1 \pmod{g(x)}.$$

In the binary case, $\eta(x) = \sigma'(x)$, thus

$$
\begin{aligned}
S(x)\sigma(x) &= \sigma'(x) \\
S(x)[\alpha^2(x) + x\beta^2(x)] &= \beta^2(x) \\
\alpha^2(x) + x\beta^2(x) &\equiv h(x)\beta^2(x) \pmod{g(x)} \\
\beta^2(x)[h(x) + x] &\equiv \alpha^2(x) \pmod{g(x)}. \quad (9.19)
\end{aligned}
$$

Let $d^2(x) = h(x) + x$. Then (9.19) can be written as

$$d^2(x)\beta^2(x) \equiv \alpha^2(x) \pmod{g(x)},$$

and thus

$$d(x)\beta(x) \equiv \alpha(x) \pmod{g(x)}. \quad (9.20)$$

Solving (9.20), i.e., finding $\alpha(x)$ and $\beta(x)$, then from (9.18) $\sigma(x)$ is obtained.

EXAMPLE 9.6 *Consider the binary irreducible $(8,2,5)$ Goppa code defined by $g(x) = x^2 + x + 1$ with $GF(2^3)$ as the extension field generated by $x^3 + x + 1$ having α as a root, i.e., $\alpha^3 = \alpha + 1$. Let the transmitted codeword be the all-zero 8-tuple and suppose that the error vector $\mathbf{e}$ is as follows:*

$$\mathbf{e} = (1,0,0,0,0,0,0,1).$$

(a) The syndrome $S(x)$ is calculated as

$$S(x) = \frac{1}{x} + \frac{1}{x - \alpha^6} \equiv \alpha^5 x + \alpha^2 \pmod{(x^2 + x + 1)}.$$

(b) The polynomial $h(x)$ is then obtained as

$$h(x) \equiv 1/S(x) \pmod{(x^2 + x + 1)},$$

i.e.,

$$h(x) = \alpha^3 x + \alpha \pmod{(x^2 + x + 1)}.$$

(c) From $d^2(x) = h(x) + x = \alpha x + \alpha \pmod{(x^2 + x + 1)}$ it is easy to obtain $d(x) = \alpha^4 x$ and then, from $d(x)\beta(x) = \alpha(x) \pmod{(x^2 + x + 1)}$, both $\beta(x) = 1$ and $\alpha(x) = \alpha^4 x$ are found.

(d) It thus follows that

$$\sigma(x) = \alpha^2(x) + x\beta^2(x) = \alpha x^2 + x,$$

which has as roots $\sigma_1 = 0$ and $\sigma_2 = \alpha^6$, i.e., 0 and α^6 are the error location numbers.

Table 9.1.   Computing $\sigma(x)$ with the Blahut algorithm.

i	$g_i(x)$	$r_i(x)$	$q_i(x)$
-1	0	$x^4 + x^2 + 1$	$\ldots$
1	1	$\alpha^3 x^3 + x^2 + \alpha^4 x + \alpha^2$	$\ldots$
2	$\alpha^4 x + \alpha$	$x^2 + \alpha^6 x + \alpha^4$	$\alpha^4 x + \alpha$
3	$\sigma(x)$	α^6	$\alpha^3 x + \alpha^6$

9.4.2   The Blahut algorithm

By using the standard decoding algorithm for a Goppa code in the binary case, with $g(x)$ over $\mathrm{GF}(2^m)$, of degree t and irreducible, this decoder can only correct up to $\lfloor t/2 \rfloor$ errors. However, it is known that, in this case, $d \geq 2t + 1$ and one should be able to correct up to t errors. Blahut (1983, p.237) observed that in this case it is still possible to perform the correction of up to t errors, using the standard decoding algorithm with no change in the code, but describing it modulo $g^2(x)$, instead of modulo $g(x)$.

EXAMPLE 9.7 *Consider again the same code, transmitted codeword and error vector of the previous example. Now however the standard decoding algorithm for Goppa codes is applied considering the results reduced modulo* $g^2(x) = x^4 + x^2 + 1$. *The resulting syndrome is as follows:*

$$S(x) = \frac{1}{x} + \frac{1}{x - \alpha^6} \equiv \alpha^3 x^3 + x^2 + \alpha^4 x + \alpha^2 \quad \mathrm{mod}\ (x^4 + x^2 + 1).$$

By using the Euclidean algorithm $\sigma(x)$ *is obtained with the help of Table 9.1 whose construction is described in (Clark and Cain 1981, pp.195–201). Finally,*

$$\sigma(x) = 1 + (\alpha^4 x + \alpha)(\alpha^3 x + \alpha^6) = x^2 + \alpha^6 x,$$

and thus, as before, the values $\sigma_1 = 0$ *and* $\sigma_2 = \alpha^6$ *are obtained.*

9.5   The asymptotic Gilbert bound

Recall that a Goppa code for which the Goppa polynomial is irreducible over $\mathrm{GF}(q^m)$ is called an irreducible Goppa code. In this section it will be shown that there exist long irreducible Goppa codes which meet the asymptotic Gilbert bound (see Appendix A).

As already seen in Subsection 5.5.3, the number of degree t polynomials, irreducible over $GF(q^m)$, is given by

$$I_t = \frac{1}{t} \sum_{d:d|t} \mu(d)(q^m)^{(t/d)},$$

where $\mu(d)$ is the Moebius function. In (Berlekamp 1968, p.80) the author gives what he called a crude lower bound on I_t, which states that

$$I_t > \frac{q^{mt}}{t} \left(1 - q^{-(mt/2)+1} \right).$$

The derivation of this bound will now proceed, i.e., it will be proven that

$$I_t = \frac{1}{t} \sum_{d:d|t} \mu(d)(q^m)^{(t/d)} > \frac{q^{mt}}{t} \left(1 - q^{-(mt/2)+1} \right).$$

By Corollary 5.6 it is known that

$$(q^m)^t = \sum_{d:d|t} dI_d,$$

where $I_d \geq 0$. Thus, it follows that

$$q^{mt} = tI_t + \sum_{d:d|t,d\neq t,d\neq 1} dI_d + I_1. \tag{9.21}$$

However, $I_1 = q^m$ and it follows from (9.21) that

$$q^{mt} \geq tI_t + q^m \tag{9.22}$$

or, equivalently,

$$I_t \leq \frac{1}{t} \left(q^{mt} - q^m \right).$$

It follows from (9.22) that $q^{mt} > tI_t$ and thus it also follows that

$$\sum_{j=0}^{t/2} q^{mj} > \sum_{j=0}^{t/2} jI_j > \sum_{d|t,d\neq t} dI_d. \tag{9.23}$$

Therefore

$$\begin{aligned} q^{mt} &= \sum_{d|t} dI_d = tI_t + \sum_{d|t,d\neq t} dI_d < tI_t + \sum_{j=0}^{t/2} jI_j \\ &< tI_t + \sum_{j=0}^{t/2} q^{mj} < tI_t + q^{(mt/2)+1}, \end{aligned}$$

where the first and second inequalities follow because of (9.23) and the third inequality used the fact that $\sum_{j=0}^{t/2} q^{mj}$ is part of a super increasing sequence, and it thus follows that

$$I_t > \frac{q^{mt}}{t}\left(1 - q^{-(mt/2)+1}\right).   \tag{9.24}$$

This derivation is similar to that in the proof of Theorem 5.10, however, employing q^m instead of q.

If $\mathbf{c}$ is a codeword with Hamming weight d, in the irreducible Goppa code with Goppa polynomial $g(x)$, then

$$\sum_{\gamma_i} \frac{c_i}{x - \gamma_i} \equiv 0 \mod g(x),$$

where $\gamma_i \in \mathrm{GF}(q^m)$ and $g(\gamma_i) \neq 0$, is a rational function which can be represented by a quotient where the numerator has degree at most $d - 1$ and the denominator has degree d. By definition, the Goppa polynomial of any Goppa code containing $\mathbf{c}$ must divide the numerator. Also, such degree t Goppa polynomials, irreducible over $\mathrm{GF}(q^m)$, are pairwise relatively prime. Therefore, the number of irreducible Goppa codes containing $\mathbf{c}$ is almost equal to the number of distinct degree t factors of the numerator, i.e., almost equal to $\lfloor (d-1)/t \rfloor$, where $\lfloor x \rfloor$ denotes the integer part of x.

The number of irreducible Goppa codes containing codewords of Hamming weight $\leq D$ will now be counted. These are called bad irreducible Goppa codes. If the number of bad irreducible Goppa codes is less than the total number of distinct irreducible Goppa codes then there must be some irreducible Goppa codes with minimum distance $d > D$.

The number of vectors of length q^m and Hamming weight d is given by

$$\binom{q^m}{d}(q-1)^d$$

and each such vector may belong to at most $\lfloor (d-1)/t \rfloor$ distinct irreducible Goppa codes. Therefore, by summing over all d, such that $d \leq D$, the total number of bad irreducible Goppa codes is obtained. By comparing this result to the lower bound for I_t it is concluded that if

$$\sum_{d=0}^{D} \left\lfloor \frac{d-1}{t} \right\rfloor (q-1)^d \binom{q^m}{d} < \frac{q^{mt}}{t}\left(1 - q^{-(mt/2)+1}\right)$$

then there exist "good" irreducible Goppa codes, i.e., Goppa codes for which $d > D$. For large $n = q^m$, by setting $t = (1-R)q^m/m$, the irreducible Goppa codes have rate $\geq R$ and the above inequality is seen to be only slightly weaker asymptotically than the Gilbert bound, i.e.,

$$\sum_{d=0}^{D} (q-1)^d \binom{n}{d} < q^{(1-R)n}.$$

Let R denote any given code rate, $0 < R < (q-1)/q$, and let $\epsilon > 0$ denote any small positive constant. It then follows that almost all irreducible q-ary Goppa codes of rate R and long block length n have minimum distance no more than ϵn less than the corresponding value provided by the Gilbert bound. From $tm = (1-R)q^m$ it follows that $R = 1 - mt/q^m$, and if $n = q^m$ and $n - k \leq mt$ then $k \geq n - mt$ and $k/n \geq 1 - mt/n = R$. Also, since

$$\sum_{d=0}^{D} \left\lfloor \frac{d-1}{t} \right\rfloor (q-1)^d \binom{q^m}{d} < \frac{q^{mt}}{t}\left(1 - q^{-(mt/2)+1}\right)$$

for large $n = q^m$ it follows that

$$\sum_{d=0}^{D} \left\lfloor \frac{d-1}{t} \right\rfloor (q-1)^d \binom{n}{d} < \frac{q^{(1-R)n}}{t}$$

or

$$\sum_{d=t+1}^{D} t \left\lfloor \frac{d-1}{t} \right\rfloor (q-1)^d \binom{n}{d} < q^{(1-R)n}.$$

9.6 Quadratic equations over GF(2^m)

The normal basis of GF(2^m) has the form

$$\gamma, \gamma^2, \ldots, \gamma^{2^{m-1}}.$$

Any element $\beta \in$ GF(2^m) can be represented in terms of the normal basis as

$$\beta = b_0\gamma + b_1\gamma^2 + \cdots + b_{m-1}\gamma^{2^{m-1}}, \tag{9.25}$$

where $b_i \in \{0,1\}$, $0 \leq i \leq m-1$, and the trace of β, denoted as $\mathrm{Tr}(\beta)$, is

$$\mathrm{Tr}(\beta) = b_0\mathrm{Tr}(\gamma) + b_1\mathrm{Tr}(\gamma^2) + \cdots + b_{m-1}\mathrm{Tr}(\gamma^{2^{m-1}})$$

or

$$\mathrm{Tr}(\beta) = b_0 + b_1 + \cdots + b_{m-1} \tag{9.26}$$

since γ is such that

$$\mathrm{Tr}(\gamma) = \mathrm{Tr}(\gamma^2) = \cdots = \mathrm{Tr}(\gamma^{2^{m-1}}) = 1.$$

Furthermore, it is noticed that

$$\beta^2 = b_0\gamma^2 + b_1\gamma^4 + \cdots + b_{m-2}\gamma^{2^{m-1}} + b_{m-1}\gamma^{2^m}$$

i.e., that

$$\beta^2 = b_{m-1}\gamma + b_0\gamma^2 + b_1\gamma^4 + \cdots + b_{m-2}\gamma^{2^{m-1}}.$$

The coefficients of β^2 are obtained simply by shifting the coefficients of β by one place to the right.

THEOREM 9.8 *The quadratic equation* $x^2 + x + \beta$, $\beta \in \mathrm{GF}(2^m)$, *has two roots in* $\mathrm{GF}(2^m)$ *if* $\mathrm{Tr}(\beta) = 0$, *and has no roots in* $\mathrm{GF}(2^m)$ *if* $Tr(\beta) = 1$. *Thus,* $f(x) = x^2 + x + \beta = (x - \lambda_1)(x - \lambda_2)$, *for* $\lambda_1, \lambda_2 \in \mathrm{GF}(2^m)$ *if* $Tr(\beta) = 0$, *but* $f(x)$ *is irreducible over* $\mathrm{GF}(2^m)$ *if* $Tr(\beta) = 1$.

Proof: Let x and x^2 be expressed in terms of the normal basis as

$$x = x_0\gamma + x_1\gamma^2 + \ldots + x_{m-1}\gamma^{2^{m-1}}$$
$$x^2 = x_{m-1}\gamma + x_0\gamma^2 + \ldots + x_{m-2}\gamma^{2^{m-1}}.$$

If x is a solution of $x^2 + x + \beta = 0$ it follows, by using (9.25) and equating coefficients of γ^{2^i}, $0 \le i \le m-1$, that

$$\begin{aligned}
x_0 + x_{m-1} &= b_0 \\
x_1 + x_0 &= b_1 \\
&\vdots \\
x_{m-1} + x_{m-2} &= b_{m-1} \qquad\qquad (+) \\
\hline
0 &= \textstyle\sum_{i=0}^{m-1} b_i = \mathrm{Tr}(\beta).
\end{aligned}$$

Thus, $\mathrm{Tr}(\beta) = 0$ is a necessary condition for $f(x) = 0$ to have a solution in $\mathrm{GF}(2^m)$. On the other hand, if $\mathrm{Tr}(\beta) = 0$ then $f(x) = 0$ will have two solutions. Namely,

$$x_0 = \lambda, \; x_1 = \lambda + b_1, \; x_2 = \lambda + b_1 + b_2, \; \ldots,$$
$$x_{m-1} = \lambda + b_1 + b_2 + \ldots + b_{m-1},$$

where $\lambda = 0$ or $\lambda = 1$. Thus $\mathrm{Tr}(\beta) = 0$ is also a sufficient condition for $x^2 + x + \beta = 0$ to have a solution in $\mathrm{GF}(2^m)$. $\square$

EXAMPLE 9.9 *(a)* $f(x) = x^2 + x + 1$, *in* $\mathrm{GF}(2)[x]$, *is irreducible because* $Tr(1) = 1$.

(b) $f(x) = x^2 + x + \alpha^2$, *in* $\mathrm{GF}(4)[x]$, *is irreducible because* $\mathrm{Tr}(\alpha^2) = \alpha^2 + \alpha^4 = 1$, *reducing modulo* $x^2 + x + 1$, *i.e., with* $\alpha^2 = \alpha + 1$.

(c) $f(x) = x^2 + x + \alpha$, in $\mathrm{GF}(8)[x]$, is reducible because $Tr(\alpha) = \alpha + \alpha^2 + \alpha^4 = 0$, with reduction modulo $x^3 + x + 1$, i.e., $\alpha^3 = \alpha + 1$. Actually, $f(x) = (x + \alpha^2)(x + \alpha^6)$.

THEOREM 9.10 *Let β be a fixed element of $\mathrm{GF}(2^m)$.*

(a) *If $\mathrm{Tr}(\beta) = 1$ then, over $\mathrm{GF}(2^m)[x]$, any irreducible quadratic $ax^2 + bx + c$ can be transformed into $A(x^2 + x + \beta)$, for some $A \in \mathrm{GF}(2^m)$, by a suitable affine transformation of variables (in $\mathrm{GF}(2^m)$).*

(b) *If $\mathrm{Tr}(\beta) = 0$ then, over $\mathrm{GF}(2^m)[x]$, any reducible quadratic $ax^2 + bx + c$ with distinct roots, can be transformed into $A(x^2 + x + \beta)$, for some $A \in \mathrm{GF}(2^m)$, by an affine transformation of variables.*

Proof:

(a) Let $f(x) = ax^2 + bx + c$ be irreducible over $\mathrm{GF}(2^m)$, with $a \neq 0$, $b \neq 0$, $c \neq 0$. Notice that if $a = 0$ then $f(x) = bx + c$ is no longer quadratic. If $b = 0$ then $f(x)$ is a perfect square, and if $c = 0$ then $f(x) = x(ax + b)$ is reducible. Replacing x by xb/a in $ax^2 + bx + c$ it follows that

$$\frac{b^2}{a}\left(x^2 + x + \frac{ac}{b^2}\right) = A(x^2 + x + d),$$

where $A = b^2/a$ and $d = ac/b^2$. By Theorem 9.8, $\mathrm{Tr}(d) = 1$. Thus,

$$\mathrm{Tr}(\beta + d) = \mathrm{Tr}(\beta) + \mathrm{Tr}(d) = 0.$$

Also, by Theorem 9.8, there exists $\lambda \in \mathrm{GF}(2^m)$ such that

$$\lambda^2 + \lambda + (\beta + d) = 0.$$

By changing $x + \lambda$ for x in $A(x^2 + x + d)$ it follows that

$$A[(x + \lambda)^2 + (x + \lambda) + d] = A[x^2 + x + (\lambda^2 + \lambda + d)] = A[x^2 + x + \beta],$$

where the last equality follows because $\lambda^2 + \lambda + d = \beta$.

(b) Let $f(x) = ax^2 + bx + c$ be reducible over $\mathrm{GF}(2^m)$, with two distinct roots. Replacing x by xb/a in $ax^2 + bx + c$ it results, as before

$$\frac{b^2}{a}\left(x^2 + x + \frac{ac}{b^2}\right) = A(x^2 + x + d),$$

where $A = b^2/a$ and $d = ac/b^2$. By Theorem 9.8, $\mathrm{Tr}(d) = 0$. Thus,

$$\mathrm{Tr}(\beta + d) = \mathrm{Tr}(\beta) + \mathrm{Tr}(d) = 0.$$

Also, by Theorem 9.8, there exists $\lambda \in GF(2^m)$ such that

$$\lambda^2 + \lambda + (\beta + d) = 0.$$

By changing $x + \lambda$ for x in $A(x^2 + x + d)$ it follows that

$$A[(x+\lambda)^2 + (x+\lambda) + d] = A[x^2 + x + (\lambda^2 + \lambda + d)] = A[x^2 + x + \beta],$$

where the last equality follows because $\lambda^2 + \lambda + d = \beta$.

$\square$

9.7   Adding an overall parity-check digit

Let $\Gamma(L, g)$ be a Goppa code over $GF(q)$, of length $n = q^m$, with $L = \{0, 1, \alpha, \ldots, \alpha^{n-2}\}$, i.e., L consists of all the elements of $GF(q^m)$, and let $g(x)$ be a polynomial of degree t with no roots in $GF(q^m)$. Let $\mathbf{c} = [c(0), c(1), \ldots, c(\alpha^{n-2})]$ be a codeword of $\Gamma(L, g)$. This $\Gamma(L, g)$ code may be extended by appending an overall parity-check digit $c(\infty)$ to each codeword, where $c(\infty)$ is given by

$$c(\infty) = - \sum_{\gamma \in GF(q^m)} c(\gamma),$$

and thus $c(\infty) \in GF(q^m)$. Equivalently,

$$\sum_{\gamma \in GF(q^m) \cup \{\infty\}} c(\gamma) = 0.$$

Recall from (9.4) that

$$\sum_{\gamma \in GF(q^m)} \frac{c(\gamma)}{x - \gamma} \equiv 0 \quad \mod g(x).$$

Adopting the convention that

$$\frac{c(\infty)}{x - \infty} = 0,$$

(9.4) can be expressed as

$$\sum_{\gamma \in GF(q^m) \cup \{\infty\}} \frac{c(\gamma)}{x - \gamma} \equiv 0 \quad \mod g(x). \tag{9.27}$$

Using the parity-check matrix $\mathbf{H}$ from (9.7) it follows that $\mathbf{c}\mathbf{H}^{\mathrm{T}} = 0$, i.e.,

$$\sum_{i=0}^{n-1} c_i h_i \gamma_i^j = 0, \ 0 \le j \le t - 1,$$

where $h_i = g^{-1}(\gamma_i)$ or, by using $\mathbf{c} = [c(0), c(1), \ldots, c(\gamma^{n-1})]$,

$$\sum_{\gamma \in \mathrm{GF}(q^m)} \frac{\gamma^i c(\gamma)}{g(\gamma)} = 0, \quad i = 0, 1, \ldots, t-1$$

or, by adding an overall parity-check digit

$$\sum_{\gamma \in \mathrm{GF}(q^m) \cup \{\infty\}} \frac{\gamma^i c(\gamma)}{g(\gamma)} = 0, \quad i = 0, 1, \ldots, t \tag{9.28}$$

with the convention that $1/\infty = 0$.

9.8   Affine transformations

DEFINITION 9.11 *An affine transformation from* $\mathrm{GF}(q^m)$ *to* $\mathrm{GF}(q^m)$ *is a transformation of the form*

$$f(x) = \frac{ax + b}{cx + d}$$

when $ad - bc \neq 0$, $a, b, c, d \in \mathrm{GF}(q^m)$.

Now we will consider the homomorphism from 2×2 matrices onto the affine transformations as follows.

$$\begin{aligned} \text{If } \sigma_1: & \quad \gamma \to (a\gamma + b)/(c\gamma + d) \\ \sigma_2: & \quad \gamma \to (e\gamma + f)/(g\gamma + h) \end{aligned}$$

then

$$\sigma_1 \sigma_2: \quad \gamma \to (i\gamma + j)/(k\gamma + l),$$

where

$$\begin{bmatrix} i & j \\ k & l \end{bmatrix} = \begin{bmatrix} a & b \\ c & d \end{bmatrix} \begin{bmatrix} e & f \\ g & h \end{bmatrix}$$

assuming $a, b, c, d \in \mathrm{GF}(q^m)$, $ad \neq bc$, and $e, f, g, h \in \mathrm{GF}(q^m)$, $eh \neq fg$, and $\gamma \in \mathrm{GF}(q^m) \cup \{\infty\}$.

Denote the elements of $\mathrm{GF}(q^m) \cup \{\infty\}$ by two-component column vectors $\begin{bmatrix} r \\ s \end{bmatrix}$, where $\begin{bmatrix} r \\ s \end{bmatrix}$ represents the elements r/s, if $s \neq 0$. $\begin{bmatrix} r \\ 0 \end{bmatrix}$ represents ∞. It is also noticed that $\begin{bmatrix} r \\ s \end{bmatrix}$ and $\alpha \begin{bmatrix} r \\ s \end{bmatrix}$, $\alpha \in \mathrm{GF}(q^m)$, represent the same element.

EXAMPLE 9.12 *Consider* $\mathrm{GF}(2^3)$, *with* α *being a primitive element satisfying* $\alpha^3 = \alpha + 1$, *and consider the affine transformation*

$$\sigma_1: \quad \gamma \to (\gamma + 1)/(\alpha^2 \gamma + \alpha^4)$$

■ *If $\gamma = \alpha^3$ then*

$$\sigma_1 \; : \quad \alpha^3 \rightarrow \quad (\alpha^3 + 1)/(\alpha^5 + \alpha^4) = \alpha/1 = \alpha$$

Using the matrix homomorphism it follows that

(a)

$$\frac{\gamma + 1}{\alpha^2 \gamma + \alpha^4} \leftrightarrow \begin{bmatrix} 1 & 1 \\ \alpha^2 & \alpha^4 \end{bmatrix}$$

(b)

$$\alpha^3 \leftrightarrow \begin{bmatrix} \alpha^3 \\ 1 \end{bmatrix}$$

(c)

$$\frac{\alpha^3 + 1}{\alpha^2 \alpha^3 + \alpha^4} \leftrightarrow \begin{bmatrix} 1 & 1 \\ \alpha^2 & \alpha^4 \end{bmatrix} \begin{bmatrix} \alpha^3 \\ 1 \end{bmatrix} = \begin{bmatrix} \alpha^3 + 1 \\ \alpha^5 + \alpha^4 \end{bmatrix} = \begin{bmatrix} \alpha \\ 1 \end{bmatrix} \leftrightarrow \alpha.$$

■ *If $\gamma = \infty$ then*

$$\infty \;\leftrightarrow\; \begin{bmatrix} 1 \\ 0 \end{bmatrix}$$

$$\frac{\infty + 1}{\alpha^2 \infty + \alpha^4} \;\leftrightarrow\; \begin{bmatrix} 1 & 1 \\ \alpha^2 & \alpha^4 \end{bmatrix} \begin{bmatrix} 1 \\ 0 \end{bmatrix} = \begin{bmatrix} 1 \\ \alpha^2 \end{bmatrix} \leftrightarrow \frac{1}{\alpha^2} = \alpha^5.$$

Recall from (9.28) that

$$\sum_{\gamma \in \mathrm{GF}(q^m) \cup \{\infty\}} \frac{\gamma^i c(\gamma)}{g(\gamma)} = 0, \quad i = 0, 1, \ldots, t.$$

If $g_1(\gamma)$ is obtained from $g(\gamma)$ by an affine transformation

$$\pi \; : \gamma \rightarrow \frac{a\gamma + b}{c\gamma + d}$$

where $a, b, c, d \in \mathrm{GF}(q^m), ad \neq bc$, and $\gamma \in \mathrm{GF}(q^m) \cup \{\infty\}$, then the code with Goppa polynomial $g_1(x)$ may be obtained from the code with Goppa polynomial $g(x)$ by the corresponding affine permutation of code digits.

9.9 Cyclic binary double-error correcting extended Goppa codes

The topic addressed here was originally developed in (Berlekamp and Moreno 1973). We consider binary Goppa codes with irreducible Goppa polynomial $g(x) = x^2 + x + \beta$, over $\mathrm{GF}(2^m)[x]$. Since by Theorem 9.10, by using a suitable affine transformation of variables in $\mathrm{GF}(2^m)$ it is possible to transform any irreducible quadratic into any other quadratic, there is only one extended double error-correcting Goppa code of length $2^m + 1$, which is the cardinality of the set $\mathrm{GF}(2^m) \cup \{\infty\}$. Therefore, the choice of $g(x) = x^2 + x + \beta$ incurs no loss of essential generality. Since $g(x)$ is irreducible, it follows from Theorem 9.8 that

$$\mathrm{Tr}(\beta) = \sum_{i=0}^{m-1} \beta^{2^i} = 1.$$

Notice that if γ is a root of $g(x)$ then $\gamma + 1$ and β/γ are also roots of $g(x)$. Hence, the code is invariant to the following permutations:

$$\begin{aligned} \rho_1 : \gamma &\;\rightarrow\; \gamma + 1 \\ \rho_2 : \gamma &\;\rightarrow\; \beta/\gamma. \end{aligned}$$

By squaring (9.27) we obtain

$$\sum_{\gamma \in \mathrm{GF}(2^m) \cup \{\infty\}} \frac{c(\gamma)}{x^2 + \gamma^2} \equiv \sum_{\gamma \in \mathrm{GF}(2^m) \cup \{\infty\}} \frac{c(\gamma)}{x + \beta + \gamma^2} \quad \mathrm{mod}\ g(x),$$

and thus the code is also invariant under

$$\rho_3 : \gamma \rightarrow \gamma^2 + \beta.$$

Notice that

$$\begin{aligned} \rho_3 : \gamma &\;\rightarrow\; \gamma^2 + \beta \\ \rho_3^2 : \gamma &\;\rightarrow\; (\gamma^2 + \beta)^2 + \beta = \gamma^4 + \beta^2 + \beta \\ \rho_3^3 : \gamma &\;\rightarrow\; (\gamma^4 + \beta^2 + \beta)^2 + \beta = \gamma^8 + \beta^4 + \beta^2 + \beta \\ &\;\;\vdots \quad\;\; \vdots \quad\; \vdots \quad\; \vdots \\ \rho_3^m : \gamma &\;\rightarrow\; \gamma^{2^m} + \mathrm{Tr}(\beta) = \gamma + 1, \end{aligned}$$

So, $\rho_3^m = \rho_1$.

LEMMA 9.13 *Extended binary Goppa codes with* $g(x) = x^2 + x + \beta, \beta \in \mathrm{GF}(2^m), \mathrm{Tr}(\beta) = 1$, *are cyclic with block length* $n = 2^m + 1$.

Proof: As remarked earlier, because of Theorem 9.10, $g(x) = x^2 + x + \beta$ causes no essential loss of generality. Suppose that λ and $\lambda + 1$ are the roots of $g(x)$, neither of which belongs to $GF(2^m)$. We show next that the permutation

$$\rho_2\rho_1 : \;\to\; \beta/(\gamma + 1), \tag{9.29}$$

obtained by applying ρ_2 first and then ρ_1, consists of a single cycle. It is necessary and sufficient to show that, for a suitable β,

$$\begin{bmatrix} 0 & \beta \\ 1 & 1 \end{bmatrix}^n \begin{bmatrix} 1 \\ 0 \end{bmatrix} = k \begin{bmatrix} 1 \\ 0 \end{bmatrix}, \quad k \in GF(2^m), \tag{9.30}$$

if and only if n is divisible by $2^m + 1$.

Any eigenvalue λ of $\begin{bmatrix} 0 & \beta \\ 1 & 1 \end{bmatrix}$ is a root of

$$\begin{vmatrix} \lambda & \beta \\ 1 & \lambda + 1 \end{vmatrix} = \lambda^2 + \lambda + \beta = 0,$$

i.e., λ must be a root of $g(x)$. Thus, from (9.30), it follows that

$$k = \lambda^n.$$

Let $\zeta \in GF(2^{2m})$, $\zeta \notin GF(2^m)$, be a primitive root of $x^{2^m+1} + 1$, i.e., $\zeta^{2^m+1} + 1 = 0$ implying $\zeta^{2^m} = \zeta^{-1}$. Also, since

$$\left(\zeta^{2^m} + \zeta\right)^{2^m} = \zeta^{2^{2m}} + \zeta^{2^m} = \zeta^{2^m} + \zeta,$$

it is concluded that $\zeta^{2^m} + \zeta = \alpha \in GF(2^m)$. It thus follows that

$$\zeta^{2^m} + \zeta + \alpha = 0$$

or

$$\zeta^{-1} + \zeta + \alpha = 0 \;\to\; \zeta^2 + \alpha\zeta + 1 = 0.$$

Let $\lambda = \zeta/\alpha$. Then, from $\zeta^2 + \alpha\zeta + 1 = 0$ we obtain $\lambda^2 + \lambda + 1/\alpha^2 = 0$. So, λ is a root of $x^2 + x + \beta$ with $\beta = 1/\alpha^2$. This is how β is selected. It follows that $\lambda^n = \zeta^n \alpha^{-n}$ and $\lambda^n \in GF(2^m)$ if and only if $\zeta^n \in GF(2^m)$, which will be the case if and only if

$$(\zeta^n)^{(2^m-1)} = 1. \tag{9.31}$$

However,

$$\zeta^{2^{2m}-1} = \zeta^{(2^m-1)(2^m+1)} = 1$$

and $\gcd(2^m - 1, 2^m + 1) = 1$, thus n must be divisible by $2^m + 1$. Finally, $2^m + 1$ is the smallest value of n for which (9.31) is satisfied and so (9.29) consists of a single cycle. □

We now consider the case when $g(x)$ is reducible in $\mathrm{GF}(2^m)[x]$.

LEMMA 9.14 *Extended binary Goppa codes with* $g(x) = x^2 + x + \beta, \beta \in \mathrm{GF}(2^m), \mathrm{Tr}(\beta) = 0$, *are cyclic with block length* $n = 2^m - 1$.

Proof: The proof begins by noticing that because of Theorem 9.10 this form of $g(x)$ causes no essential loss of generality. Suppose that λ and $\lambda + 1$ are the roots of $g(x)$ over $\mathrm{GF}(2^m)$. Consider the permutation

$$\rho_2\rho_1 \ : \ \gamma \to \frac{\beta}{\gamma + 1} \tag{9.32}$$

where it is noticed that

$$\lambda \to \frac{\beta}{\lambda + 1} = \lambda$$

and

$$1 + \lambda \to \frac{\beta}{\lambda} = 1 + \lambda,$$

i.e., the roots of $g(x)$ are fixed by $\rho_2\rho_1$. It will now be proven that the permutation $\rho_2\rho_1$ consists of a single cycle. Just as was done before, a necessary and sufficient condition for that to happen is that

$$\begin{bmatrix} 0 & \beta \\ 1 & 1 \end{bmatrix}^n \begin{bmatrix} 1 \\ 0 \end{bmatrix} = k \begin{bmatrix} 1 \\ 0 \end{bmatrix}, \quad k \in \mathrm{GF}(2^m)$$

for a suitable β, if and only n is divisible by $2^m - 1$. The eigenvalues λ of $\begin{bmatrix} 0 & \beta \\ 1 & 1 \end{bmatrix}$ are the roots of $g(x)$, $\lambda^2 + \lambda + \beta = 0$. As before, let ζ be a primitive root of $x^{2^m+1} + 1$, $\zeta \in \mathrm{GF}(2^{2m})$, and it is noticed that

$$\left(\zeta^{2^m} + \zeta \right)^{2^m} = \zeta^{2^{2m}} + \zeta^{2^m} = \zeta^{2^m} + \zeta = \alpha \ \in \mathrm{GF}(2^m).$$

Thus, $\zeta^{2^m} + \zeta + \alpha = 0$ with $\zeta^{2^m} = \zeta^{-1}$ leads to $\zeta^2 + \alpha\zeta + 1 = 0$ and by making $\lambda = \zeta/\alpha$, we obtain $\lambda^2 + \lambda + 1/\alpha^2 = 0$. So, λ is a root of $\lambda^2 + \lambda + \beta = 0$ with $\beta = 1/\alpha^2$. Now, consider

$$\lambda^n = \zeta^n \alpha^{-n}$$

or

$$\lambda^{n(2^m+1)} = \alpha^{-n(2^m+1)}, \tag{9.33}$$

since $\zeta^{2^m+1} = 1$, and it follows from (9.33) that

$$\lambda^{2n} = \alpha^{-2n}$$

because $\lambda^{2^m-1} = 1$ and $\alpha^{2^m-1} = 1$. Thus either

(a) $\lambda = \alpha^{-1}$, which contradicts $\lambda^2 + \lambda + 1/\alpha^2 = 0$

    or

(b) $\lambda^{2n} = \alpha^{-2n} = 1$ which, since $\gcd(2, 2^m - 1) = 1$, implies $(2^m - 1)|n$. Finally, $2^m - 1$ is the least value of n for which (9.33) holds and thus (9.32) consists of a single cycle.

$\square$

9.10   Extending the Patterson algorithm for decoding Goppa codes

It is now considered the decoding of binary Goppa codes using the Patterson algorithm, when the Goppa polynomial, of degree t, is a product of non-repeated irreducible factors over GF(2^m). In order to succeed one must be able to find, as before, $\alpha(x), \beta(x) \in$ GF(2^m)[x], $\deg(\alpha(x)) \le t/2$, $\deg(\beta(x)) \le (t-1)/2$, to obtain

$$\sigma(x) = \alpha^2(x) + x\beta^2(x).$$

Start with the key-equation

$$S(x)\sigma(x) \equiv \sigma'(x) \quad \mod g(x),$$

where it is known that $\gcd(\sigma(x), \sigma'(x)) = 1$. Suppose that $\gcd(S(x), g(x)) = a(x)$. It follows that

$$\sigma'(x) = \beta^2(x) = a^2(x)b^2(x).$$

Factoring $a(x)$ out in the key-equation, it follows that

$$S_1(x)\sigma(x) \equiv a(x)b^2(x) \quad \mod g_1(x)$$

where $S(x) = a(x)S_1(x)$ and $g(x) = a(x)g_1(x)$. Next it is required to find $h(x)$ such that

$$S_1(x)h(x) \equiv 1 \quad \mod g_1(x).$$

From

$$S_1(x)\sigma(x) \quad \equiv \quad a(x)b^2(x) \quad \mod g_1(x)$$
$$\text{and}$$
$$S_1(x)h(x) \quad \equiv \quad 1 \quad \mod g_1(x)$$

it follows that
$$\sigma(x) \equiv h(x)a(x)b^2(x) \mod g_1(x)$$

i.e.,

$$\begin{aligned}
\alpha^2(x) + x\beta^2(x) &\equiv h(x)a(x)b^2(x) \mod g_1(x) \\
\alpha^2(x) + xa^2(x)b^2(x) &\equiv h(x)a(x)b^2(x) \mod g_1(x) \\
\alpha^2(x) &\equiv b^2(x)[h(x)a(x) + xa^2(x)] \mod g_1(x) \\
\alpha^2(x) &\equiv b^2(x)d^2(x) \mod g_1(x).
\end{aligned}$$

Now, it is required to solve the congruence

$$\alpha(x) \equiv b(x)d(x) \mod g_1(x)$$

to obtain $\alpha(x)$ and $\beta(x)$. Notice that this congruence can be solved if

$$\deg \alpha(x) + \deg b(x) < \deg g_1(x)$$

or

$$\deg \alpha(x) + [\deg \beta(x) - \deg a(x)] < \deg g(x) - \deg a(x)$$

i.e.,

$$\deg \alpha(x) + \deg \beta(x) < \deg g(x),$$

which condition is satisfied by the initial assumption on the degrees of $\alpha(x)$ and $\beta(x)$. Thus, having $\alpha(x)$ and $b(x)$ such that $\alpha(x) \equiv b(x)d(x) \mod g_1(x)$ we obtain

$$\sigma(x) = \alpha^2(x) + xa^2(x)b^2(x) = \alpha^2(x) + x\beta^2(x).$$

It is noticed that if $a(x) = 1$ then $\gcd[S(x), g(x)] = 1$, and thus not only irreducible Goppa codes but in fact any binary Goppa code, with Goppa polynomial of degree at most 2, can be decoded by the original Patterson algorithm. This follows because, for any Goppa code $\gcd[\sigma(x), g(x)] = 1$, and if

$$\sigma(x) = \sigma_2 x^2 + \sigma_1 x + \sigma_0$$

then $\sigma'(x) = \sigma_1$ is a scalar and one can always find $h(x)$ such that $S(x)h(x) \equiv 1 \mod g(x)$.

9.11 Problems with solutions

(1) Construct the parity-check matrix for the binary Goppa code with Goppa polynomial $g(x) = x$ in GF(8)$[x]$.

Solution: Let α be a primitive element of GF(8). One can use $L = \{1, \alpha, \alpha^2, \ldots \alpha^6\}$ as the locator numbers since $g(\alpha^i) = \alpha^i \neq$

$0,\ 0 \le i \le 6$. By definition, the code is characterized by

$$\sum_{i=0}^{6} \frac{c_i}{x - \alpha^i} \equiv 0 \quad \mathrm{mod}\ x, \quad c_i \in \{0, 1, \}.$$

However, the $x - \alpha^i, 0 \le i \le 6$, are units in the ring of polynomials modulo x, and thus it follows that

$$\sum_{i=0}^{6} \frac{c_i}{x - \alpha^i} \equiv \sum_{i=0}^{6} \frac{c_i}{\alpha^i} \equiv \sum_{i=0}^{6} c_i \alpha^{-i}$$

and thus

$$\mathbf{H} = [1\ \ \alpha^{-1}\ \ \alpha^{-2}\ \ \alpha^{-3}\ \ \alpha^{-4}\ \ \alpha^{-5}\ \ \alpha^{-6}],$$

or

$$\mathbf{H} = [1\ \ \alpha^6\ \ \alpha^5\ \ \alpha^4\ \ \alpha^3\ \ \alpha^2\ \ \alpha]$$

is the parity-check matrix. If $\alpha^3 = \alpha + 1$ then, over GF(2), $\mathbf{H}$ is written as

$$\mathbf{H} = \begin{bmatrix} 1 & 1 & 1 & 0 & 1 & 0 & 0 \\ 0 & 0 & 1 & 1 & 1 & 0 & 1 \\ 0 & 1 & 1 & 1 & 0 & 1 & 0 \end{bmatrix}$$

which is the parity-check matrix of a binary $(7, 4, 3)$ Hamming code.

(2) Construct the parity-check matrix for the binary Goppa code with Goppa polynomial $g(x) = x^2$ in GF(8)$[x]$.

Solution: Using the parity-check matrix in the form given in (9.7), with $t = 2$, it follows that

$$\mathbf{H} = \begin{bmatrix} h_0 & h_1 & h_2 & h_3 & h_4 & h_5 & h_6 \\ h_0\gamma_0 & h_1\gamma_1 & h_2\gamma_2 & h_3\gamma_3 & h_4\gamma_4 & h_5\gamma_5 & h_6\gamma_6 \end{bmatrix}$$

where $h_i = g^{-1}(\alpha^i)$ and $\gamma_i = \alpha^i$, thus $\mathbf{H}$ can be written as

$$\mathbf{H} = \begin{bmatrix} 1 & \alpha^5 & \alpha^3 & \alpha & \alpha^6 & \alpha^4 & \alpha^2 \\ 1 & \alpha^6 & \alpha^5 & \alpha^4 & \alpha^3 & \alpha^2 & \alpha \end{bmatrix}$$

If $\alpha^3 = \alpha + 1$ then, over GF(2), $\mathbf{H}$ is written as

$$\mathbf{H} = \begin{bmatrix} 1 & 1 & 1 & 0 & 1 & 0 & 0 \\ 0 & 1 & 1 & 1 & 0 & 1 & 0 \\ 0 & 1 & 0 & 0 & 1 & 1 & 1 \\ 1 & 1 & 1 & 0 & 1 & 0 & 0 \\ 0 & 0 & 1 & 1 & 1 & 0 & 1 \\ 0 & 1 & 1 & 1 & 0 & 1 & 0 \end{bmatrix}.$$

Notice that **H** has rank three over GF(2) because row 1 equals row 4, row 2 equals row 6, and the modulo 2 element-by-element modulo 2 addition of rows 2 and 3 equals row 5. Thus, as expected, the code obtained is the same as that in the earlier problem.

(3) Use the Patterson algorithm to decode the code in Problem 1 when the received polynomial is $r(x) = x^2$, i.e., $\mathbf{r} = (0,0,1,0,0,0,0)$.

Solution: The syndrome is computed as

$$S(x) \equiv \sum_{i=0}^{6} \frac{r_i}{x - \alpha^i} \quad \text{mod } g(x)$$

$$S(x) \equiv \frac{1}{x - \alpha^2} \quad \text{mod } x$$

or

$$S(x) \equiv \frac{1}{-\alpha^2} \equiv \alpha^5 \quad \text{mod } x.$$

Thus, $S(x) \equiv \alpha^5 \mod x$ and from $S(x)h(x) \equiv 1 \mod x$ one gets $h(x) \equiv \alpha^2 \mod x$. The key-equation is

$$S(x)\sigma(x) \equiv \sigma'(x) \quad \text{mod } g(x)$$

and since one writes $\sigma^2(x) = \alpha^2(x) + x\beta^2(x)$, it follows that

$$S(x)[\alpha^2(x) + x\beta^2(x)] \equiv \beta^2(x) \quad \text{mod } g(x)$$

or

$$\alpha^2(x) + x\beta^2(x) \equiv \beta^2(x)h(x) \quad \text{mod } g(x)$$

i.e.

$$\alpha^2(x) \equiv d^2(x)\beta^2(x) \quad \text{mod } g(x),$$

where $d^2(x) \equiv h(x) + x \mod g(x)$.
Remark: If $h(x) \equiv x \mod g(x)$ then from $S(x)h(x) \equiv 1 \mod g(x)$, i.e., $S(x)x \equiv 1 \mod g(x)$, one sets $\sigma(x) = x$ and $\sigma'(x) = 1$, and terminates.

In this problem one has

$$d^2(x) \equiv h(x) + x = \alpha^2 + x \equiv \alpha^2 \quad \text{mod } x$$

thus $d(x) \equiv \alpha \mod x$ is taken into $\alpha(x) \equiv \beta(x)d(x) \mod g(x)$, and $\alpha\beta(x) \equiv \alpha(x) \mod g(x)$ has a solution $\alpha(x) = \alpha$, $\beta(x) = 1$ producing $\sigma(x) = \alpha^2 + x$, which leads to the correct decoding of $r(x)$.

(4) Use the Blahut algorithm to decode the received polynomial in Problem 3.

Solution: The syndrome is computed as

$$S(x) \equiv \sum_{i=0}^{6} \frac{r_i}{x - \alpha^i} \mod g^2(x)$$

$$S(x) \equiv \frac{1}{x + \alpha^2} \mod x^2$$

or

$$S(x) \equiv \alpha^3 x + \alpha^5 \mod x^2,$$

and solving the key-equation

$$S(x)\sigma(x) \equiv \sigma'(x) \mod x^2$$

one obtains $\sigma(x) = \alpha^4 x + \alpha^6$, which has α^2 as its single root.

(5) Apply the Patterson algorithm to decode the extended Goppa code with $g(x) = x(x+1) = x^2 + x$ in GF(8)$[x]$, when the received n-tuple is $\mathbf{r} = (0, 1, 0, 0, 0, 1, 0)$ and $L = \{\infty, \alpha, \alpha^2, \alpha^3, \alpha^4, \alpha^5, \alpha^6\}$, where ∞ stands for the location of the overall parity-check digit.

Solution: The syndrome is computed as

$$S(x) \equiv \frac{1}{x - \alpha} + \frac{1}{x - \alpha^5} \equiv \alpha^2 x + 1 \mod x^2 + x,$$

obtained by using $\alpha^3 = \alpha + 1$, α being primitive in GF(8). Thus $h(x)$ can be found from $S(x)h(x) \equiv 1 \mod x^2 + x$ as

$$h(x) \equiv \alpha^3 x + 1$$

and

$$d^2(x) = h(x) + x \equiv \alpha x + 1 \mod x^2 + x$$

leads to

$$d(x) \equiv \alpha^4 x + 1 \mod x^2 + x.$$

It follows from $\alpha(x) \equiv d(x)\beta(x) \mod x^2 + x$ that

$$\alpha(x) \equiv (\alpha^4 x + 1)\beta(x) \equiv (\alpha x + 1)\beta(x) \mod x^2 + x,$$

which is solved by $\beta(x) = 1$, $\alpha(x) = \alpha^4 x + 1$. Thus, $\sigma(x) = \alpha^2(x) + x\beta^2(x) = \alpha x^2 + x + 1$ has roots α and α^5, and $\mathbf{r}$ is correctly decoded as the all-zero codeword. Why did it work, since $g(x)$ is *reducible*?

Chapter 10

CODING-BASED CRYPTOSYSTEMS

10.1 Introduction

In (Shannon 1949) it was explicitly stated that the problem of designing good cryptosystems is basically equivalent to finding difficult problems. A cryptosystem may be constructed in a manner that breaking it is equivalent to solving a problem known to be hard to solve. In (McEliece 1978) the hard to solve problem selected to build a cryptosystem was that of decoding a general linear code. In this chapter, we look at cryptosystems which employ error-correcting codes in their construction.

10.2 McEliece's public-key cryptosystem

McEliece (1978) introduced a public-key cryptosystem of the block cipher-type based on algebraic coding theory. The security of this system relics on the computational complexity of solving the general decoding problem for linear codes. This system presents the intended receiver with an easy to solve problem which is the decoding of a t error-correcting Goppa code, for which there is a fast decoding algorithm. Intruders, however, are faced with the far more difficult task of decoding a linear t error-correcting code, for which there is no known fast decoding algorithm.

10.2.1 Description of the cryptosystem

In the following we assume binary irreducible Goppa codes are used. As already seen, for any irreducible polynomial $g(x)$ over $GF(2^m)$, of degree t and any integer n, $1 \leq t < n \leq 2^m$, there exists a binary irreducible Goppa code $\Gamma(L, g)$ of length n and dimension k, $k \geq n - mt$, that can correct any t or fewer errors. The code block length n is determined by the size of the set L which is chosen as a subset of $\mathrm{GF}(2^m)$. We have also seen an algebraic (fast) decoding algorithm for such codes (see Section 9.4). To set up this cryptosystem we choose integers t and n, $1 \leq t < n \leq 2^m$, and then randomly select a monic irreducible polynomial $g(x)$ of degree t over $\mathrm{GF}(2^m)$. The probability of obtaining a monic irreducible polynomial of degree t over $\mathrm{GF}(2^m)$ in this way is given by the ratio of the favorable cases N_t to the total number of possible choices 2^{mt}, i.e., $N_t/2^{mt} = 2^{-mt} \sum_{d|t} \mu(t/d) 2^{md}$. Once a polynomial is chosen, it can be tested for irreducibility by applying the factorization techniques previously studied.

A binary t error-correcting irreducible Goppa code $\Gamma(L, g)$ of block length n and dimension k, $k \geq n - mt$ has a binary $(n - k) \times n$ parity-check matrix $\mathbf{H}$. The code $k \times n$ generator matrix $\mathbf{G}$ is obtained from $\mathbf{H}$. This generator matrix $\mathbf{G}$ is then transformed by the operation $\mathbf{SGP}$, i.e., by multiplying the matrices $\mathbf{S}$, $\mathbf{G}$, and $\mathbf{P}$, in that order, where $\mathbf{S}$ denotes a binary $k \times k$ invertible (dense) matrix and $\mathbf{P}$ denotes an $n \times n$ permutation matrix. As will become clearer when we cover the deciphering operation, the role of $\mathbf{S}$ is to scramble the data bits. Remember that a permutation matrix has a single 1 per row and per column. Obviously the identity matrix satisfies this definition but performs no permutation of coordinates. We call $\mathbf{G}'$ the matrix resulting from the product $\mathbf{SGP}$. It is easy to see that $\mathbf{G}'$ generates a binary linear code equivalent to the original Goppa code, i.e., a code with the same block length, same dimension and same minimum distance as the original Goppa code. The matrix $\mathbf{G}'$ is made public, e.g., available from a directory, while matrices $\mathbf{S}, \mathbf{G}$, and $\mathbf{P}$ are kept secret.

10.2.2   Encryption

At the sender's side, encryption proceeds by generating enciphered n-tuples as follows. The binary message is segmented into blocks of k bits. Each k-bit block is then multiplied by $\mathbf{G}'$. The generated n-tuple is added to a random error pattern of weight t. The resulting n-tuple is the enciphered message, ready for transmission or storage.

10.2.3   Decryption

On receiving an enciphered n-tuple, the intended receiver first performs on it an inverse permutation $\mathbf{P}^{-1}$ in order to restore the Goppa code. Notice that what results, after the application of $\mathbf{P}^{-1}$, is a codeword of the Goppa code, containing t random errors. The decoding algorithm for binary irreducible Goppa codes produces a scrambled k-tuple

$\mathbf{m}'$. Finally, the message is recovered by multiplying $\mathbf{m}'$ by $\mathbf{S}^{-1}$, i.e., $\mathbf{m} = \mathbf{m}'\mathbf{S}^{-1}$, where $\mathbf{S}^{-1}$ denotes the inverse of the scrambling matrix $\mathbf{S}$.

10.2.4   Cryptanalysis

The possibilities for an intruder to break this system are either by finding $\mathbf{G}$ from $\mathbf{G}'$, or by finding $\mathbf{m}$ directly from the enciphered n-tuple, or by using a secret (so far unpublished) fast decoding algorithm for linear binary codes. The parameters suggested by McEliece were $n = 2^{10} = 1024$ and $t = 50$. Notice that $\mathbf{G}'$ is combinatorially equivalent to $\mathbf{G}$. Also, the number of codes which are combinatorially equivalent to a given code is extremely large for $n \geq 100$ and $k \geq 50$. Therefore, an attempt to factor $\mathbf{G}'$ into $\mathbf{SGP}$ is bound to fail. There are $\cong 10^{149}$ possible irreducible Goppa polynomials of degree 50 over $\mathrm{GF}(2^{10})$. The suggested code dimension was 524 in order to make it unfeasible both syndrome decoding and exhaustive search of all codewords. Later, it was shown that the code parameters can be chosen in a way that increases the cryptanalytic complexity of the system and decreases its data expansion (Adams and Meijer 1989).

In the sequel, we will determine the optimum number of errors to be added, to yield high security against a certain type of attack also to be described. We will show that with high probability there is essentially only one inverse transformation from $\mathbf{G}'$ to an easily decodable Goppa code, i.e., the inverse transformation known to the receiver. Let $\mathbf{c} = \mathbf{m}\mathbf{G}' + \mathbf{e}$ be the enciphered message n-tuple, where $\mathbf{m}$ is the cleartext k-tuple, $\mathbf{G}'$ is the modified Goppa code generator matrix and $\mathbf{e}$ is a weight t random error pattern. Since $\mathbf{m}$ is a k-tuple, a possible attack is as follows. Choose k bits from $\mathbf{c}$, i.e.,

$$\mathbf{c}_k = \mathbf{m}\mathbf{G}'_k + \mathbf{e}_k, \tag{10.1}$$

where $\mathbf{c}_k = (c_{i_1}, c_{i_2}, \ldots, c_{i_k})$, i.e., $\mathbf{c}_k$ denotes any k components of $\mathbf{c}$, $\mathbf{e}_k$ denotes the corresponding k components of $\mathbf{e}$ and $\mathbf{G}'_k$ is the square matrix formed with columns $i_1, i_2, \ldots, i_k$ of $\mathbf{G}'$. From (10.1) we can write

$$\mathbf{c}_k + \mathbf{e}_k = \mathbf{m}\mathbf{G}'_k,$$

or, if $\mathbf{G}'_k$ is invertible

$$(\mathbf{c}_k + \mathbf{e}_k)(\mathbf{G}'_k)^{-1} = \mathbf{m}. \tag{10.2}$$

If the k components of $\mathbf{e}_k$ are all-zero then (10.2) gives $\mathbf{m}$ directly. This means that recovery of $\mathbf{m}$ is possible without conventional decoding. We now look at the computational effort necessary to succeed in such an attack. The formula for computing the probability p_k of choosing k zero components for $\mathbf{e}$ is

$$p_k = \frac{\dbinom{n-t}{k}}{\dbinom{n}{k}},$$

because $\mathbf{e}$ has t 1's and $n - t$ 0's and the choices are made without replacement. An intruder must perform on average $1/p_k$ attempts to be successful. Each such attempt involves the inversion of the matrix $\mathbf{G}'$, which is estimated to require k^α, $2 \leq \alpha \leq 3$ steps. Therefore, the approximate number of steps required on average is

$$k^\alpha \frac{\dbinom{n}{k}}{\dbinom{n-t}{k}}. \tag{10.3}$$

By exhaustive search (Adams and Meijer 1989) it has been found that (10.3) has a maximum for $t = 37$, which is $\cong 2^{84.1}$. The value of (10.3) for $t = 50$ is $\cong 2^{80.7}$. As a consequence of using a lower value of t, an increased value for the dimension k was obtained, namely 654, compared with the old value 524. The resulting higher code rate means in cryptographic terms lower data expansion.

10.2.5   Trapdoors

We will now estimate the likelihood of existing various distinct transforms of the public-key $\mathbf{G}'$ into an easily decodable code. Let $\mathbf{F}$ define an equivalence relation on the set of $k \times n$ matrices of full rank as $\mathbf{XFY}$ if and only if there exists an invertible $k \times k$ matrix $\mathbf{S}$ and an $n \times n$ permutation matrix $\mathbf{P}$ such that $\mathbf{X} = \mathbf{SYP}$. In this way, by calling $[\mathbf{X}]$ the equivalence class induced by $\mathbf{F}$ containing $\mathbf{X}$, then the private matrix $\mathbf{G}$ is in the equivalence class of the public matrix $\mathbf{G}'$. We are interested in the answer for the following question. Are there other equivalent Goppa code generator matrices in the equivalence class $[\mathbf{G}']$? Let us suppose that, over the set of all full rank $k \times n$ matrices, the Goppa code generator matrices are probabilistically distributed according to a uniform probability distribution. Then it follows that the expected number, E, of Goppa code generator matrices in an equivalence class $\mathbf{F}$ is given by

$$E = |\mathbf{G}|/|\mathbf{C}|,$$

where $|\mathbf{G}|$ denotes the number of Goppa code generator matrices, for given n and k, and $|\mathbf{C}|$ denotes the number of equivalence classes in $\mathbf{F}$. In the article (Adams and Meijer 1989) it is reasoned that, for $t = 50$ and $k = 524$, $|\mathbf{G}|$ is not greater than the number of irreducible polynomials of degree 50 over $\mathrm{GF}(1024)$, and that $|\mathbf{C}|$ is approximated by the ratio between the number of full rank 524×1024 matrices divided by the size of an equivalence class. The value obtained was

$$E < 2^{504}/2^{500000}. \tag{10.4}$$

For $t = 37$ and $k = 654$, an even smaller E results. Provided the uniform distribution assumption holds, the bound given in (10.4) indicates that for practical purposes $\mathbf{G}$ is the only Goppa code generator matrix contained in the equivalence class of $\mathbf{G}'$.

10.3 Secret-key algebraic coding systems

Secret-key algebraic coding schemes, introduced by Rao and Nam (1989), are similar to McEliece's, however, they keep the generator matrix $\mathbf{G}'$ private, as well as the $\mathbf{S}$, $\mathbf{G}$ and $\mathbf{P}$ matrices. Their interesting aspect is the use of simpler error-correcting codes, thus requiring a relatively lower computational overhead compared to their public-key counterpart.

10.3.1 A (possible) known-plaintext attack

Try to independently solve matrices for each possible column vector of $\mathbf{G}'$. This requires a large number of $\mathbf{m}, \mathbf{c}$ pairs. This attack can be countered by a periodic change of the keys by the sender.

10.3.2 A chosen-plaintext attack

A chosen-plaintext attack essentially involves two steps.

(1) Find $\mathbf{G}'$ using a large set of chosen $\mathbf{m}$ and associated $\mathbf{c}$, i.e., chosen $\mathbf{m}, \mathbf{c}$ pairs.

(2) Then find unknown messages $\mathbf{m}$ from the intercepted $\mathbf{c}$, using $\mathbf{G}'$ found in step 1.

Notice that step 2 is exactly the same as that faced by a cryptanalist in the public-key algebraic coding cryptosystem. The cryptanalist can feed the message $\mathbf{m} = (0, 0, 0, \ldots, 0, 1, 0, \ldots, 0)$ containing a single 1 in the ith position. Since t is much smaller than n, most of the bits in the corresponding ciphertext $\mathbf{c}$ are those of $\mathbf{g}'_i$. We can then use a majority logic decoding rule to estimate $\mathbf{g}'_i$ as follows. Let $\mathbf{c}$ represent one such estimate of $\mathbf{g}'_i$. Then, for a fixed i, repeat this procedure and obtain various estimates of $\mathbf{g}'_i$. By taking a majority vote among these estimates we obtain the correct value of $\mathbf{g}'_i$, and thus find out $\mathbf{g}'_i$, the ith row of $\mathbf{G}'$. If the system does not allow such a message, we can then feed plaintexts m_1 and m_2 differing only in the ith position, and thus $m_1 - m_2 = (0,0,0,\ldots,0,1,0,\ldots,0)$. Thus, by subtracting the associated ciphertexts $\mathbf{c}_1$ and $\mathbf{c}_2$, and because the code is linear, we obtain

$$\mathbf{c}_1 - \mathbf{c}_2 = \mathbf{g}'_i + \mathbf{e}_1 - \mathbf{e}_2,$$

where the weight of $\mathbf{e}_1 - \mathbf{e}_2$ is at most $2t$. Since $2t$ is much smaller than n, most of the bits in $\mathbf{c}_1 - \mathbf{c}_2$ are those of $\mathbf{g}'_i$. Then, as described earlier, we can use a majority logic decoding rule to estimate $\mathbf{g}'_i$. By making $i = 1, 2, \ldots, k$, all rows of $\mathbf{G}'$ can be thus be obtained. Having obtained $\mathbf{G}'$ we can proceed to step 2. If t is small, the work factor (average number of steps) required in step 2 will be relatively small. Notice that the majority rule attack works because the ratio $t/n \ll 0.5$.

10.3.3　　A modified scheme

In this scheme (Rao and Nam 1989) it is proposed to use higher Hamming weight error patterns, aiming at ratios $t/n \approx 0.5$, to foil the previous chosen-plaintext attack. This serves as an example of the fact that cryptosystems based on computational complexity are always subject to threats represented by the enemy finding short cuts to the apparently difficult to solve problems.

Encryption
Using the already defined notation, encryption consists of the following steps:

(1) Form the product $\mathbf{G}' = \mathbf{SG}$.

(2) Multiply $\mathbf{m}$ by $\mathbf{G}'$ and add a noise vector $\mathbf{e}$ to the resulting $\mathbf{mG}'$.

(3) Multiply $(\mathbf{mG}'+\mathbf{e})$ by $\mathbf{P}$ and obtain the ciphertext $\mathbf{C} = (\mathbf{mG}'+\mathbf{e})\mathbf{P}$.

The error patterns are now chosen from the standard array (see Section 2.4) for the code employed, one error pattern from each coset, each error pattern having weight $\cong n/2$. The total number of allowable error patterns is thus 2^{n-k} since this is the total number of cosets. The receiver has to store in advance the table of pre-chosen error patterns. This can represent a practical difficulty. Notice that the permutation $\mathbf{P}$ was applied to $\mathbf{mG}' + \mathbf{e}$. This has the effect of permuting (scrambling) the error pattern. However, this system has no randomness associated with it.

Decryption

(1) Start with $\mathbf{c} = \mathbf{m}\mathbf{G'}\mathbf{P} + \mathbf{e}\mathbf{P}$.

(2) Apply $\mathbf{P}^{-1}$ to $\mathbf{c}$, i.e.,

$$\mathbf{c'} = \mathbf{c}\mathbf{P}^{-1} = \mathbf{m'}\mathbf{G} + \mathbf{e}.$$

(3) Recover $\mathbf{m'}$ as follows. First find $\mathbf{e}$ by calculating the associated syndrome as

$$\mathbf{c'}\mathbf{H}^{\mathrm{T}} = \mathbf{m'}\mathbf{G}\mathbf{H}^{\mathrm{T}} + \mathbf{e}\mathbf{H}^{\mathrm{T}} = \mathbf{e}\mathbf{H}^{\mathrm{T}},$$

and then using it to locate the error $\mathbf{e}$ in the error table. Recover $\mathbf{m'}$ by correcting the error pattern $\mathbf{e}$.

(4) Finally, $\mathbf{m} = \mathbf{m'}\mathbf{S}^{-1}$.

Cryptanalysis

We consider the following a chosen-plaintext attack. Rewrite $\mathbf{c}$ as follows:

$$\mathbf{c} = (\mathbf{m}\mathbf{S}\mathbf{G} + \mathbf{e})\mathbf{P} = \mathbf{m}\mathbf{S}\mathbf{G}\mathbf{P} + \mathbf{e}\mathbf{P} = \mathbf{m}\mathbf{G''} + \mathbf{e}\mathbf{P},$$

where $\mathbf{G''} = \mathbf{S}\mathbf{G}\mathbf{P} = [\mathbf{g}_i'']$ for $i = 1, 2, \ldots, k$, and $\mathbf{g}_i''$ is the ith row vector of $\mathbf{G''}$.

LEMMA 10.1 *There are at least $k!$ combinatorially equivalent $(n, k, 3)$ linear codes.*

LEMMA 10.2 *The number N_s of non-singular $k \times k$ matrices $\mathbf{S}$ over* $GF(2)$ *is given by*

$$N_s = \prod_{i=0}^{k-1}(2^k - 2^i)$$

and is lower-bounded as $N_s > 2^{k^2 - k}$.

The two lemmas given earlier serve to show the difficulties associated with attacks aiming at finding $\mathbf{S}, \mathbf{G}$ and $\mathbf{P}$. Let $\mathbf{c}_j$ and $\mathbf{c}_l$ denote two different ciphertexts that were extracted for the same chosen-plaintext $\mathbf{m}$. Then it follows that

$$\begin{aligned}
\mathbf{c}_j &= \mathbf{m}\mathbf{G''} + \mathbf{e}_j\mathbf{P} \\
\mathbf{c}_l &= \mathbf{m}\mathbf{G''} + \mathbf{e}_l\mathbf{P},
\end{aligned}$$

and thus

$$\mathbf{c}_j - \mathbf{c}_l = (\mathbf{e}_j - \mathbf{e}_l)\mathbf{P}.$$

This last step produces one value for $(\mathbf{e}_j - \mathbf{e}_l)\mathbf{P}$. It is repeated until all possible pairs of error patterns are exhausted. The total number of distinct pairs of error patterns is $2^{n-k}(2^{n-k} - 1)/2 \cong 2^{2(n-k)-1}$. Using a pair of messages differing only in the ith position, we obtain (as we did previously)

$$\mathbf{c}_1 - \mathbf{c}_2 = \mathbf{g}_i'' + (\mathbf{e}_1 - \mathbf{e}_2)\mathbf{P},$$

and

$$\mathbf{g}_i'' = \mathbf{c}_1 - \mathbf{c}_2 - (\mathbf{e}_1 - \mathbf{e}_2)\mathbf{P}.$$

However, the correctness of the estimates of $\mathbf{g}_i''$ cannot be tested independently. This means that the complete $\mathbf{G}''$ matrix has to be obtained and verified. The work factor involved in this task is

$$T \geq (1/2)(2^{2(n-k)-1})^k,$$

i.e., an exponential function of both k and $n-k$. Thus this system can be considered secure against the type of chosen-plaintext attack described.

10.4   Problems with solutions

(1) What is the meaning of *breaking* a cryptosystem?

     Solution: According to (Massey 1998), breaking a cryptographic system means that the enemy cryptanalyst is able, either always or unacceptably often, to do the very thing that the system is intended to prevent, i.e., to recover the plaintext from the enciphered transmission or to trick the receiver into accepting a fraudulent message.

(2) In (Krouk 1993) the following public-key cypher based on error-correcting codes was proposed. Consider the n-dimensional vector space over $\mathrm{GF}(q)$. Let $\mathcal{E}$ denote a set of q-ary vectors of length n, where the first l coordinates can be any l-tuple from $\mathrm{GF}(q)$ and the remaining $n - l$ coordinates are zeros. Let $\mathcal{C}$ be an (n, k) linear block code over $\mathrm{GF}(q)$, with generator matrix $\mathbf{G}$, capable of efficient correction of errors represented by the elements in $\mathcal{E}$. Assume $\mathbf{Q}$ and $\mathbf{P}$ are, respectively, $l \times n$ and $(n - l) \times n$ matrices over $\mathrm{GF}(q)$ and suppose that $\mathbf{M}$ is an $n \times n$ matrix of the form

$$\mathbf{M} = \begin{bmatrix} \mathbf{Q} \\ \mathbf{P} \end{bmatrix}$$

which has a multiplicative inverse $\mathbf{M}'$ in $\mathrm{GF}(q)$, i.e., $\mathbf{M}'\mathbf{M}^{\mathrm{T}} = \mathbf{I}_n$, the $n \times n$ identity matrix. Let $\mathbf{J}$ be the matrix defined as

$$\mathbf{J} = \begin{bmatrix} \mathbf{Q} \\ \mathbf{I}_{n-l} \end{bmatrix}$$

and notice that $\mathbf{eM} = \mathbf{eJ}$. The set $\mathcal{E}$ and matrices $\mathbf{Q}$ and $\mathbf{G'} = \mathbf{GM}$ constitute the public-key. Matrices $\mathbf{G}$ and $\mathbf{M'}$ constitute the private-key. The ciphertext consists of the n-tuple $\mathbf{y} = \mathbf{uG'} + \mathbf{eJ} = \mathbf{uG'} + \mathbf{eM}$, where $\mathbf{u}$ is a k-tuple message and $\mathbf{e} \in \mathcal{E}$ is an error vector. Decryption by the intended receiver consists of computing $\mathbf{r} = \mathbf{y}(\mathbf{M'})^{\mathrm{T}} = (\mathbf{uG'} + \mathbf{eM})(\mathbf{M'})^{\mathrm{T}} = \mathbf{uG} + \mathbf{e}$. A decoder for code $\mathcal{C}$ is employed to correct the error $\mathbf{e}$ and produce the codeword $\mathbf{c} = \mathbf{uG}$. Finally, $\mathbf{u}$ is extracted from the codeword $\mathbf{c}$. Devise a way to break Krouk's public-key cipher.

Solution: The reader will find a detailed solution to this problem in (Rocha Jr. and Macedo 1996).

(3) Prove Lemma 10.1.

Solution: The $\mathbf{G}$ matrix can be written in reduced echelon form (see Section 2.2) as $\mathbf{G} = [\mathbf{I}_k | \mathbf{g}]$ where $\mathbf{I}_k$ is a $k \times k$ identity matrix and $\mathbf{g}$ is a $k \times (n - k)$ matrix. There are $k!$ row permutations of the matrix $\mathbf{g}$, all of which produce $(n, k, 3)$ codes. All of these code generator matrices can, by row permutation and column permutation, be obtained from $\mathbf{G}$. Therefore they are all combinatorially equivalent to $\mathbf{G}$.

(4) Prove Lemma 10.2.

Solution: Let us consider the construction of non-singular $\mathbf{S}$ matrices. The number of choices for the first row is $2^k - 1$, i.e., any nonzero binary k-tuple will do. The second row must be linearly independent from the first row. In binary this means just that they are nonzero and distinct. Therefore, there are $2^k - 2$ choices for the second row. The chosen two rows, by linear combination, give rise to $2^2 = 4$ linearly dependent rows. Thus, there are $2^k - 2^2$ choices for the third row. Continuing in this way, we find that there are $2^k - 2^{i-1}$ choices for the ith row and the equality in the lemma follows. The smallest term in the k-term product is $2^k - 2^{k-1}$ and thus $N_o > (2^k - 2^{k-1})^k = 2^{k^2-k}$.

Chapter 11

MAJORITY LOGIC DECODING

11.1 Introduction

Majority logic decoding is a well-established branch of the theory of error-correcting codes. It draws heavily from the theory of finite geometries, e.g., Euclidean and projective geometries, both for the construction of codes and for their decoding. In general the resulting codes are not very powerful but their decoders are both simple and very fast. For practical applications most attention is given to either cyclic block codes or to convolutional codes of rate $1/n$ or $(n-1)/n$ because of the resulting simplified decoders.

11.2 One-step majority logic decoding

Any (n, k) linear block code $\mathcal{C}$, over $\mathrm{GF}(q)$, has a parity-check matrix $\mathbf{H}$ and the codewords $\mathbf{c}, \mathbf{c} \in \mathcal{C}$, satisfy $\mathbf{c}\mathbf{H}^{\mathrm{T}} = 0$. Denoting the jth row of $\mathbf{H}$ by $\mathbf{h}_j$ the following parity-check equation can be written

$$\mathbf{c}\mathbf{h}_j^{\mathrm{T}} = \sum_{i=0}^{n-1} h_{ji}c_i = 0.$$

By linear combinations of the rows of $\mathbf{H}$, up to q^{n-k} parity-check equations can be formed. In the sequel conditions are derived which are to be satisfied by the parity-check equations to be used for majority logic decoding.

DEFINITION 11.1 *Given a (n, k) linear block code $\mathcal{C}$, with codewords* $\mathbf{c} = (c_0, c_1, \ldots, c_{n-1})$, $c_i \in \mathrm{GF}(q)$, $0 \leq i \leq n-1$, *a set of parity-check equations in $\mathcal{C}$ is orthogonal on coordinate i if c_i appears in every* *parity-check equation of the set, and every c_j, $j \neq i$, appears in at most one parity-check equation of the set.*

THEOREM 11.2 *If a set of J parity-check equations in C is orthogonal on coordinate i, then c_i can be correctly decoded, provided that at most $\lfloor J/2 \rfloor$ errors occurred in the received n-tuple.*

Proof: Let $\mathbf{r} = \mathbf{c} + \mathbf{e}$, where $\mathbf{r}$ denotes the received n-tuple, $\mathbf{c}$ denotes the transmitted codeword and $\mathbf{e}$ denotes the error pattern n-tuple, or the noise n-tuple. Since, by definition, c_i appears in all J parity-check equations, each parity-check equation in the set can be divided by the corresponding h_{ji}, $h_{ji} \neq 0$, i.e.,

$$
\begin{aligned}
A_j &= h_{ji}^{-1} \sum_{l=0}^{n-1} h_{jl} r_l = h_{ji}^{-1} \sum_{l=0}^{n-1} h_{jl} e_l \\
&= e_i + \sum_{l, l \neq i} h_{ji}^{-1} h_{jl} e_l.
\end{aligned}
$$

(a) If $e_i = 0$ then the at most $\lfloor J/2 \rfloor$ errors will affect at most $\lfloor J/2 \rfloor$ parity-check equations and thus at least $J - \lfloor J/2 \rfloor$ parity-check equations, i.e., at least half of the parity-check equations are equal to $e_i = 0$.

(b) If $e_i \neq 0$ then the other, at most, $\lfloor J/2 \rfloor - 1$ errors will affect at most $\lfloor J/2 \rfloor - 1$ parity-check equations and thus $J - \lfloor J/2 \rfloor + 1$ parity-check equations, i.e., more than half of the parity-check equations (parity sums) will equal e_i.

Summarizing, we can correctly find e_i by a majority vote over the A_j.   □

COROLLARY 11.3 *If for any i, $0 \leq i \leq n - 1$, J parity-check equations orthogonal on a coordinate i can be constructed, then the code can correct $\lfloor J/2 \rfloor$ errors.*

The decoding procedure described in Theorem 11.2 is called *one-step majority logic decoding*. The number of errors that can be corrected by majority logic in one-step is denoted by $t_{\mathrm{ML}} = \lfloor J/2 \rfloor$. As a consequence it follows that the code minimum distance $d_{\min}$ satisfies: $d_{\min} \geq 2t_{ML} + 1$ and thus one-step majority logic decoding is efficient if $\lfloor J/2 \rfloor$ is close to $\lfloor (d_{\min} - 1)/2 \rfloor$.

DEFINITION 11.4 *If $J = d_{\min} - 1$ parity-check sums orthogonal on e_i, $1 \leq i \leq n - 1$, can be formed, then the code is said to be completely orthogonalizable in one step.*

Let $\mathbf{H}$ denote a parity-check matrix of an (n, k) code $\mathcal{C}$ over $\mathrm{GF}(q)$ and let $\mathbf{w}_1, \mathbf{w}_2, \ldots, \mathbf{w}_J$ be a set of J orthogonal vectors in the row space of $\mathbf{H}$. It follows that any $\mathbf{w}_i$, $1 \leq i \leq J$, can be expressed as a linear combination of rows of $\mathbf{H}$, where a row of $\mathbf{H}$ is denoted by $\mathbf{h}_i$, $0 \leq i \leq n - k - 1$, i.e.,

$$\mathbf{w}_j = (w_{j0}, w_{j1}, \ldots, w_{j,n-1}) = a_{j0}\mathbf{h}_0 + a_{j1}\mathbf{h}_1 + \cdots + a_{j,n-k-1}\mathbf{h}_{n-k-1},$$

where $a_{ji} \in \mathrm{GF}(q)$, $0 \leq i \leq n - k - 1$.

If $\mathbf{H}$ is in reduced echelon form then it follows that

$$w_{j0} = a_{j0}, \; w_{j1} = a_{j1}, \ldots, w_{j,n-k-1} = a_{j,n-k-1}.$$

Let $\mathbf{r} = (r_0, r_1, \ldots, r_{n-1})$ denote a received vector. The syndrome $\mathbf{S}$ associated with $\mathbf{r}$ is given by

$$\mathbf{S} = (s_0, s_1, \ldots, s_{n-k-1}) = \mathbf{r}\mathbf{H}^{\mathrm{T}},$$

where $s_i = \mathbf{r}\mathbf{h}_i^{\mathrm{T}}$, $0 \leq i \leq n - k - 1$.

Consider next the parity-check sums A_j, $1 \leq j \leq J$, defined by the dot product between $\mathbf{w}_j$ and $\mathbf{r}$, i.e., $A_j = \mathbf{w}_j\mathbf{r}$.

$$\begin{aligned} A_j &= (a_{j0}\mathbf{h}_0 + a_{j1}\mathbf{h}_1 + \cdots + a_{j,n-k-1}\mathbf{h}_{n-k-1})\mathbf{r} \\ &= a_{j0}\mathbf{h}_0\mathbf{r} + a_{j1}\mathbf{h}_1\mathbf{r} + \cdots + a_{j,n-k-1}\mathbf{h}_{n-k-1}\mathbf{r} \\ &= a_{j0}s_0 + a_{j1}s_1 + \cdots + a_{j,n-k-1}s_{n-k-1}. \end{aligned}$$

It follows that the check-sums A_j can be written as a linear combination of syndrome digits having for coefficients the first $n - k$ digits of the $\mathbf{w}_j$ orthogonal vectors, $1 \leq j \leq J$.

11.3 Multiple-step majority logic decoding I

DEFINITION 11.5 *A set of J parity-check equations is orthogonal on the set of coordinates $i_1, i_2, \ldots, i_r$ if for some $\mathrm{GF}(q)$ coefficients $A_1, A_2, \ldots, A_r$ the sum $A_1 c_{i_1} + A_2 c_{i_2} + \cdots + A_r c_{i_r}$ appears in every parity-check equation of the set and every c_i, $i \notin \{i_1, i_2, \ldots, i_r\}$, appears in at most one of the J parity-check equations of the set.*

The following theorems establish the error correction capability of one-step and of multi-step majority logic decoding.

THEOREM 11.6 *Let $\mathcal{C}$ be an (n, k) code, over $\mathrm{GF}(q)$, whose dual code has minimum distance $\overline{d}$. Then the number, denoted by J, of parity-check sums orthogonal on an error digit is upper bounded by $J \leq \left\lfloor \frac{n-1}{\overline{d}-1} \right\rfloor$.*

Proof: Suppose there exist J vectors in the dual code of $\mathcal{C}$ which are orthogonal on a given coordinate. Each of these J vectors has at least

$\bar{d} - 1$ nonzero components in the remaining $n - 1$ coordinates, i.e., a total of at least $J(\bar{d} - 1)$ nonzero coordinates which are distinct because of the orthogonality assumption. The quantity $J(\bar{d} - 1)$ cannot exceed $n - 1$ and the theorem thus follows. □

THEOREM 11.7 *Let C be an (n, k) code, over $\mathrm{GF}(q)$, whose dual code has minimum distance $\bar{d}$. Then the number, denoted by J, of parity-check sums orthogonal on a set of digits is upper bounded by $J \leq \left\lfloor \frac{2n}{\bar{d}} \right\rfloor - 1$.*

Proof: Suppose there exist J vectors in the dual code of C which are orthogonal on a set B of coordinates, containing b components. Excluding the b components in the set B, let a_l denote the number of other digits checked at the lth parity-check equation. Since these equations correspond to codewords in the dual code, which has minimum distance $\bar{d}$, it follows that

$$b + a_l \geq \bar{d}.$$

Summing these J equations it follows that

$$Jb + \sum_{l=1}^{J} a_l \geq J\bar{d}.$$

Because of the orthogonality condition in the set of J parity-check equations it follows that

$$\sum_{l=1}^{J} a_l \leq n - b.$$

Eliminating b by combining the last two inequalities the result is

$$Jn + \sum_{l=1}^{J} a_l \geq J \left(\bar{d} + \sum_{l=1}^{J} a_l \right). \tag{11.1}$$

Another condition is now derived to allow the elimination of $\sum_{l=1}^{J} a_l$ in (11.1). Subtracting two codewords orthogonal on the set B of coordinates produces another codeword containing zeros in the b places from B, having Hamming weight at least $\bar{d}$, i.e.,

$$a_l + a_{l'} \geq \bar{d}, \text{ for } l \neq l'.$$

The number of such distinct pairs of orthogonal equations is $J(J-1)/2$ and each a_l appears in $J - 1$ such pairs. Adding all such equations gives

$$(J - 1) \sum_{l=1}^{J} a_l \geq \frac{J(J-1)}{2}\bar{d}$$

which allows the elimination of $\sum_{l=1}^{J} a_l$ in (11.1) to produce $J \leq \left\lfloor \frac{2n}{d} \right\rfloor - 1.$

□

11.4   Multiple-step majority logic decoding II

Assume J check sums orthogonal on a sum of B digits and that the dual code minimum distance $\bar{d}$ is odd. Let a_l, $l = 1, 2, \ldots, J$, denote the number of other digits checked by the sum B in the lth parity-check equation, i.e., the lth check sum. Since each check sum corresponds to a codeword in the dual code, it follows that

$$B + a_l \geq \bar{d}. \tag{11.2}$$

Due to the orthogonality condition among the J check sums, it follows that

$$n - B \geq \sum_{l=1}^{J} a_l. \tag{11.3}$$

Also, by summing the J inequalities given by (11.2) produces

$$JB + \sum_{l=1}^{J} a_l \geq J\bar{d}. \tag{11.4}$$

Combining (11.3) and (11.4) to eliminate B, it follows that

$$Jn \geq (J - 1) \sum_{l=1}^{J} a_l + J\bar{d}. \tag{11.5}$$

If $\bar{d}$ is odd then at most one check sum could check only $(\bar{d} - 1)/2$ or fewer other digits. All other check sums must check at least $(\bar{d} + 1)/2$ other digits. Pairing the check sum for which $a_l \leq (\bar{d} - 1)/2$ with the remaining $J - 1$ check sums, and then pairing those $J - 1$ check sums for which $a_l \geq (\bar{d} + 1)/2$ gives, for $a_l \neq a_l'$,

$$a_l + a_{l'} \geq \begin{cases} \bar{d}, & (J - 1) \text{ times} \\ \bar{d} + 1, & (J - 1)(J - 2)/2 \text{ times}. \end{cases}$$

Since each a_l appears in $J - 1$ of the sums $a_l + a_{l'}$, it follows that

$$(J - 1) \sum_{l=1}^{J} a_l \geq (J - 1)\bar{d} + \frac{(J - 1)(J - 2)}{2}(\bar{d} + 1). \tag{11.6}$$

Eliminating $\sum_{l=1}^{J} a_l$ between (11.5) and (11.6) it follows that

$$J \leq 2 \left(\frac{n+2}{\bar{d}+1} \right) - 1.$$

11.5   Reed–Muller codes

Reed–Muller (RM) codes are binary linear codes. For given positive integers m and r, $r < m$, there is an RM code of block length 2^m called an rth order RM code of length 2^m.

The generator matrix of RM codes is described in nonsystematic form, because it is convenient for decoding. To construct the generator matrix of RM codes the following definition is required.

DEFINITION 11.8 *Given two vectors* $\mathbf{a} = (a_0, a_1, \ldots, a_{n-1})$ *and* $\mathbf{b} = (b_0, b_1, \ldots, b_{n-1})$, *the Hadamard product* $\mathbf{ab}$ *is defined as*

$$\mathbf{ab} = (a_0 b_0, a_1 b_1, \ldots, a_{n-1} b_{n-1}),$$

i.e., the components of $\mathbf{ab}$ *are obtained by the componentwise multiplication of the corresponding components in* $\mathbf{a}$ *and* $\mathbf{b}$.

The generator matrix for the rth order RM code of length 2^m is the $k \times 2^m$ matrix $\mathbf{G}$ defined as

$$\mathbf{G} = \begin{bmatrix} \mathbf{G}_0 \\ \mathbf{G}_1 \\ \vdots \\ \mathbf{G}_r \end{bmatrix},$$

where $\mathbf{G}_0$ denotes the all 1's 2^m-tuple; $\mathbf{G}_1$ is an $m \times 2^m$ matrix where the columns are all distinct m-tuples, and $\mathbf{G}_i$, $2 \leq i \leq r$, is a matrix with rows consisting of all i-fold distinct Hadamard products of rows of $\mathbf{G}_1$. It is immediate to check that the rows of $\mathbf{G}$ are linearly independent and it thus follows that

$$k = 1 + \binom{m}{1} + \binom{m}{2} + \cdots + \binom{m}{r} = \sum_{i=0}^{r} \binom{m}{i}$$

$$n - k = 2^m - k = \sum_{j=0}^{m-r-1} \binom{m}{j}.$$

Also, every row of $\mathbf{G}_l$ has Hamming weight 2^{m-l} which is an even number. Therefore, the resulting code contains only codewords of even Hamming weight, because linear combinations of binary vectors with even Hamming weight must have an even Hamming weight. Because the rows of $\mathbf{G}_r$ have Hamming weight 2^{m-r}, the code minimum distance is at most 2^{m-r}.

By using the Reed decoding algorithm (Peterson and Weldon Jr. 1972, p.316), i.e., multiple-step orthogonalization, up to $(1/2)2^{m-r} - 1$ errors can be corrected in any codeword. Thus the code minimum distance is at least $2^{m-r} - 1$, but because all codewords have even Hamming weight it follows that $d_{\min} = 2^{m-r}$. In connection with RM codes it is worth reading (Massey 1963).

11.6   Affine permutations and code construction

Cyclic codes are known to be invariant under a cyclic shift. A cyclic shift is also called a cyclic permutation. There are cyclic codes which are invariant under other permutations.

Let $\mathcal{C}$ be a cyclic code of block length $n = q^m - 1$ generated by the polynomial $g(x)$. Let $\mathcal{C}_e$ be the code of length q^m obtained by extending $\mathcal{C}$, by appending an overall parity-check denoted by c_∞, where

$$c_\infty = -(c_0 + c_1 + \cdots + c_{n-1}).$$

Consider the coordinates in a codeword vector labeled with elements of $\mathrm{GF}(q^m)$. The nonzero elements of $\mathrm{GF}(q^m)$ are represented by α^i, $0 \le i \le q^m - 1$, where α is a primitive element in the multiplicative group of $\mathrm{GF}(q^m)$. The zero element in $\mathrm{GF}(q^m)$ will be denoted by α^∞. A vector $\mathbf{c} = (c_{n-1}, \ldots, c_1, c_0, c_\infty)$ in $\mathcal{C}_e$ will have component c_i at location α^i.

DEFINITION 11.9 *A group of permutations is called transitive if, for every pair of locations (X, Y) in a codeword, there is a permutation in the group which sends X to Y, possibly rearranging other locations too.*

DEFINITION 11.10 *A group of permutations is called doubly transitive if, for every two pairs of locations (X_1, Y_1) and (X_2, Y_2) with $X_1 \ne X_2$ and $Y_1 \ne Y_2$, in an n-tuple, $n = q^m$, there is a permutation in the group which simultaneously sends X_1 to Y_1 and sends X_2 to Y_2, possibly rearranging other locations too.*

DEFINITION 11.11 *An affine permutation is a permutation which takes the component at location X in an n-tuple to location $aX + b$, $a \ne 0$, where a and b are any fixed elements in $\mathrm{GF}(q^m)$.*

The set of all affine permutations forms a group under composition because

(1) If location X goes to location $Y = aX + b$ and location Y goes to location $Z = a'X + b'$ then X goes to $Z = aa'X + a'b + b = AX + B$, and

(2) The inverse of $Y = aX + b$ is the permutation $a^{-1}X - a^{-1}b$.

The group of affine permutations is doubly transitive because the pair of equations

$$\begin{array}{rcl} Y_1 & = & aX_1 + b \text{ associated with the pair } (X_1, Y_1) \\ Y_2 & = & aX_2 + b \text{ associated with the pair } (X_2, Y_2) \end{array}$$

has a unique solution for given a and b, for $X_1 \neq X_2, Y_1 \neq Y_2$.

THEOREM 11.12 *Any code of block length* $n = q^m$ *that is invariant under the group of affine permutations can be made into a cyclic code by dropping the location at* α^∞.

Proof: Let α be the primitive element of $\mathrm{GF}(q^m)$, used to indicate the location numbers. The permutation $Y = \alpha X$ is an affine permutation and performs a cyclic shift for the nonzero locations, keeping α^∞ fixed. By dropping the position at α^∞ the theorem then follows. $\qquad\square$

DEFINITION 11.13 *Let* n *and* k *be integers with respective radix-q representation*

$$\begin{array}{rcl} n & = & n_0 + n_1q + n_2q^2 + \cdots + n_{m-1}q^{m-1}, \ 0 \leq n_i < q \\ k & = & k_0 + k_1q + k_2q^2 + \cdots + k_{m-1}q^{m-1}, \ 0 \leq k_i < q. \end{array}$$

The integer k *is called a radix-q descendent of* n *if* $k_i \leq n_i$, $i = 0, 1, \ldots, m - 1$.

In general there is no alternative but to use Definition 11.13 to determine whether a given integer k is a radix-q descendent of an integer n. For the special case where q is a prime number p the following theorem is applicable.

THEOREM 11.14 (LUCAS'S THEOREM) *Let* p *be a prime number and let* n *and* k *be two positive integers expressed in their radix-p expansion as*

$$\begin{array}{rcl} n & = & n_0 + n_1p + n_2p^2 + \cdots + n_{m-1}p^{m-1}, \\ k & = & k_0 + k_1p + k_2p^2 + \cdots + k_{m-1}p^{m-1}, \end{array}$$

then

$$\binom{n}{k} \equiv \prod_{i=0}^{m-1} \binom{n_i}{k_i} \quad \bmod p$$

and $\binom{n}{k} \equiv 0$ *modulo* p *if and only if* k *is not a radix-p descendent of* n.

Proof: The proof follows by expanding $(1+x)^n$ in two different ways and equating the coefficients of equal powers of x. Since p is a prime number it follows that

$$(1+x)^{p^i} = 1 + x^{p^i} \quad \text{mod } p.$$

For an arbitrary n it follows that

$$
\begin{aligned}
(1+x)^n &= (1+x)^{n_0 + n_1 p + \cdots + n_{m-1} p^{m-1}} \\
&= (1+x)^{n_0} (1+x)^{n_1 p} \cdots (1+x)^{n_{m-1} p^{m-1}}. \quad (11.7)
\end{aligned}
$$

The binomial expansion is next applied to both sides of (11.7) to produce

$$
\sum_{k=0}^{n} \binom{n}{k} x^k = \left[\sum_{k_0=0}^{n_0} \binom{n_0}{k_0} x^{k_0} \right] \left[\sum_{k_1=0}^{n_1} \binom{n_1}{k_1} x^{k_1} \right] \cdots \left[\sum_{k_{m-1}=0}^{n_{m-1}} \binom{n_{m-1}}{k_{m-1}} x^{k_{m-1}} \right]
$$

and equating the coefficients of x^k on both sides it follows that

$$
\binom{n}{k} = \sum_{u} \binom{n_0}{k_0} \binom{n_1}{k_1} \cdots \binom{n_{m-1}}{k_{m-1}} \quad (11.8)
$$

where u runs over all m-tuples $(k_0, k_1, \ldots, k_{m-1})$ having $k_i < p$, $0 \le i \le m-1$, such that

$$k = k_0 + k_1 p + k_2 p^2 + \cdots + k_{m-1} p^{m-1},$$

which is recognized as the radix-p expansion of k, which is unique. Thus there is only one value for u to satisfy the equality in (11.8) and the sum reduces to

$$
\binom{n}{k} = \prod_{i=0}^{m-1} \binom{n_i}{k_i} \quad \text{mod } p.
$$

The last part of the theorem follows because $\binom{n_i}{k_i} = 0$ if $k_i > n_i$, i.e., the case when k is not a radix-p descendent of n. $\qquad \square$

Next the characterization will be given of cyclic codes that can be extended to a code which is invariant under the group of affine permutations.

THEOREM 11.15 *Let α be a primitive element of $\mathrm{GF}(q^m)$, where q is a power of a prime number p. Let C be a cyclic code of block length $q^m - 1$ generated by a polynomial $g(x)$, with $g(1) \neq 0$, and let C_e be the code obtained by extending C with an overall parity-check. The code C_e is invariant under the group of affine permutations if and only if whenever α^k is a root of $g(x)$ then, for all the nonzero k' radix-p descendent of k, $\alpha^{k'}$ is also a root of $g(x)$.*

Proof: Let $X' = aX + b$ denote an affine permutation. Let $X_1, X_2, \ldots,$ X_s denote the location numbers of the nonzero components of a codeword $\mathbf{c}$ and let $Y_1, Y_2, \ldots, Y_s$ denote the values of these components. Also, let $X'_1, X'_2, \ldots, X'_s$ denote the location numbers of the nonzero components under the affine permutation, i.e., $X'_i = aX_i + b$, $1 \leq i \leq s$.

(a) Suppose that if $g(\alpha^k) = 0$ then $g(\alpha^{k'}) = 0$, where $k' \neq 0$ is a radix-p descendent of k. It is next shown that the permutation of a codeword is also a codeword. It is known that $c(x) = m(x)g(x)$ and that

$$c_\infty = -(c_0 + c_1 + \cdots + c_{n-1}).$$

Also, since $c(x) = \sum_{i=0}^{n-1} c_i x^i$ it follows that

$$c(\alpha^j) = \sum_{i=0}^{n-1} c_i \alpha^{ij} = \sum_{l=1}^{s} Y_l X_l^j = 0 \tag{11.9}$$

for every j such that $g(\alpha^j) = 0$. If desired, the symbol at location α^∞ could be included in the sum in (11.9) because it contributes with zero to the sum.

For the permuted vector $\mathbf{c}' = (c'_0, c'_1, \ldots, c'_{n-1}, c'_\infty)$ to be a codeword it is required that $c'_\infty + c'_{n-1} + \cdots + c'_1 + c'_0 = 0$, which is obviously true after the permutation, and that

$$\sum_{l=1}^{s} Y_l (X'_l)^j = \sum_{l=1}^{s} Y_l (aX_l + b)^j$$

$$= \sum_{l=1}^{s} Y_l \sum_{k=0}^{j} \binom{j}{k} a^k X_l^k b^{j-k}$$

$$= \sum_{k=0}^{j} \binom{j}{k} c(\alpha^k) a^k b^{j-k}.$$

Now, $\binom{j}{k} \equiv 0$ modulo p unless k is a radix-p descendent of j, in which case $c(\alpha^k) = 0$ by hypothesis. Thus the permuted vector $\mathbf{c}'$ is a codeword. Next the converse will be proved.

(b) Assume that C_e is invariant under the group of affine permutations. Then every codeword satisfies

$$\sum_{l=1}^{s} Y_l (aX_l + b)^j = 0,$$

for every a and b, and every j such that $g(\alpha^j) = 0$ and, as seen earlier, it follows that

$$\sum_{k=0}^{j} \binom{j}{k} c(\alpha^k) a^k b^{j-k} = 0. \tag{11.10}$$

Denoting by K the number of radix-p descendents of j and calling them $k_l, l = 1, 2, \ldots, K$, then (11.10) can be written as

$$\sum_{l=0}^{K} \binom{j}{k_l} c(\alpha^{k_l}) a^{k_l} b^{j-k_l} = 0. \tag{11.11}$$

Since $\binom{j}{k_l} \not\equiv 0$ modulo p and the sum in (11.11) is equal to zero for arbitrary a and b, it is concluded that $c(\alpha^{k_l}) = 0$, because there are $(q^m - 1)^2$ nonzero choices for pairs (a, b) and only K, $K < j < q^m$, values for $c(\alpha^{k_l})$, and each such pair (a, b) leads to a distinct permutation.

$\square$

11.7 A class of one-step decodable codes

This section describes a class of cyclic codes over $GF(q)$ of length $n = JL$, $L \neq 0 \mod p$, $q = p^s$, where p is a prime number and s is a positive integer, which uses affine permutations and is one-step majority logic decodable. The block length n is chosen as a primitive value which is composite, i.e.,

$$n = q^m - 1 = JL.$$

It follows that

$$x^n - 1 = (x^J - 1)(1 + x^J + x^{2J} + \cdots + x^{(L-1)J})$$

and let

$$a(x) = 1 + x^J + x^{2J} + \cdots + x^{(L-1)J}.$$

It is assumed throughout that $x^n - 1 = g(x)h(x)$ and that $h^\perp(x)$ denotes the reciprocal polynomial of $h(x)$. The nonzero elements of $GF(q^m)$ are either zeros of $x^J - 1$ or zeros of $a(x)$, but not both. Let α be a primitive element of GF(q^m). Then α^L is an element of order J, and the roots of $x^J - 1$ are $1, \alpha^L, \alpha^{2L}, \ldots, \alpha^{(J-1)L}$. The roots of $a(x)$ can be written as α^j, where j is not a multiple of L. Let $h^{\perp}(x)$ denote a dual code generator polynomial whose roots are $\alpha^j, 1 \le j \le q^m - 1$, where j or a radix-q descendent of j is not a multiple of L. It follows that $a(x)$ is a multiple of $h^{\perp}(x)$. It is noticed that the code generated by $h^{\perp}(x)$ is cyclic and, because of Lemma 11.16, it can be extended to a code with the doubly transitive invariant property.

LEMMA 11.16 *Let* $S = \{1, 2, \ldots, q^m - 1\}$ *be a set of integers where* $q = p^s$, *p is a prime number and s is a positive integer. Let* S_1 *be a set obtained from* S, *after deleting from* S *all elements which are multiples of some integer* $L, L \ne 0$ *modulo* p, $1 \le L \le q^m - 1$. *If* $j \in S_1$ *then its conjugates, i.e.,* jq^i *modulo* $q^m - 1$, $1 \le i \le m - 1$, *also belong to* S_1.

Proof: Since $j \in S_1$ is not a multiple of L, and $L \ne 0$ modulo p, it follows that no jq^i, where $q = p^s$, can be a multiple of L and thus all jq^i modulo $q^m - 1$, $1 \le i \le m - 1$, belong to S_1. $\square$

Notice that the codewords $a(x), xa(x), \ldots, x^{J-1}a(x)$ each has Hamming weight L and it is easy to verify that no two of them have a nonzero element in common. Next an overall parity-check will be added to each one of these codewords. The result is J codewords of the extended code, orthogonal on the extended symbol, which takes on the same value in all J codewords and is nonzero because $L \ne 0$ modulo p. Next these J codewords are divided by the value of the extended symbol. Therefore the value at location α^{∞} becomes 1 in all J codewords. By applying the affine permutation $Y = \alpha X + \alpha^{n-1}$ to these J codewords of the extended code the result is J codewords with a 1 in location α^{n-1}. By dropping the symbol at location α^{∞} the result is a set of J codewords in the cyclic code generated by $h^{\perp}(x)$ which are orthogonal on position α^{n-1}. The following theorem has been established.

THEOREM 11.17 *Let* $n = q^m - 1 = JL$ *and let* $\mathcal{C}$ *be a cyclic code over* GF(q) *of block length* n, *whose dual code* $\mathcal{C}^{\perp}$ *has* $h^{\perp}(x)$ *for its generator polynomial. The roots of* $h^{\perp}(x)$ *are specified by* α^j, $1 \le j \le q^m - 1$, *where* j *or a radix-q descendent of* j *is not a multiple of* L. *Then* $\mathcal{C}$ *is majority logic decodable and has a minimum distance of at least* J.

The roots of the generator polynomial $g(x)$ for the cyclic code $\mathcal{C}$ are characterized as follows. The polynomial $g(x)$ has α^{-j} as a root whenever j or a radix-q descendent of j is a multiple of L, including $j = 0$. This result follows by noticing that, since $\mathcal{C}$ and $\mathcal{C}^{\perp}$ are dual, α^{-j} is a root of $g(x)$ whenever α^j is not a root of $h^{\perp}(x)$.

EXAMPLE 11.18 *Let $q = 2$ and let $n = 2^4 - 1 = 15 = 3 \times 5$. It follows that*

$$x^{15} - 1 = (x^5 - 1)(x^{10} + x^5 + 1),$$

where $J = 5$ and $L = 3$. Let α be primitive in $GF(2^4)$ such that $\alpha^4 = \alpha + 1$. The following set of extended codewords of $C^{\perp}$ are derived from $x^{10} + x^5 + 1$, namely codewords c_1, c_2, c_3, c_4 and c_5.

Location	∞	0	1	2	3	4	5	6	7	8	9	10	11	12	13	14
c_1	1	1	0	0	0	0	1	0	0	0	0	1	0	0	0	0
c_2	1	0	1	0	0	0	0	1	0	0	0	0	1	0	0	0
c_3	1	0	0	1	0	0	0	0	1	0	0	0	0	1	0	0
c_4	1	0	0	0	1	0	0	0	0	1	0	0	0	0	1	0
c_5	1	0	0	0	0	1	0	0	0	0	1	0	0	0	0	1

The roots of $x^5 - 1$ are $1, \alpha^3, \alpha^6, \alpha^9, \alpha^{12}$ and the roots of $a(x) = x^{10} + x^5 + 1$ are $\alpha, \alpha^2, \alpha^4, \alpha^5, \alpha^7, \alpha^8, \alpha^{10}, \alpha^{11}, \alpha^{13}, \alpha^{14}$. It follows that the roots of $h^{\perp}(x)$ are $\alpha, \alpha^2, \alpha^4, \alpha^8, \alpha^5, \alpha^{10}$ and the roots of $g(x)$ are $1, \alpha^{-3}, \alpha^{-6}, \alpha^{-7}, \alpha^{-9}, \alpha^{-11}, \alpha^{-12}, \alpha^{-13}, \alpha^{-14}$, i.e., $1, \alpha, \alpha^2, \alpha^3, \alpha^4, \alpha^6, \alpha^8, \alpha^9, \alpha^{12}$. It follows that

$$
\begin{aligned}
g(x) &= (x+1)(x^4 + x + 1)(x^4 + x^3 + x^2 + x + 1) \\
&= x^9 + x^6 + x^5 + x^4 + x + 1.
\end{aligned}
$$

Since $d \geq J + 1$ and $J = 5$ it follows that $d \geq 6$. However, since the Hamming weight of $g(x)$ is 6 it is concluded that $d = 6$.

11.8   Generalized Reed–Muller codes

The generalized Reed–Muller (GRM) codes over $GF(q)$ constitute a class of codes containing subclasses which are majority logic decodable. The GRM codes are defined by extending a cyclic GRM code. Only cyclic GRM codes of primitive block length will be considered in the sequel. The GRM codes are obtained from the cyclic GRM codes by appending an overall parity-check symbol to the codewords. Let α be a primitive element in the multiplicative group of $GF(q^m)$.

DEFINITION 11.19 *Let s be a nonnegative integer with radix-q expansion*

$$s = s_0 + s_1 q + s_2 q^2 + \cdots + s_{m-1} x^{m-1}.$$

The radix-q weight of s is defined as

$$W_q(s) = \sum_{i=0}^{m-1} s_i.$$

DEFINITION 11.20 *The cyclic GRM code of order r and block length $n = q^m - 1$ over $\mathrm{GF}(q)$ is the cyclic code with generator polynomial $g(x)$ having α^j for its roots, $1 \le j \le q^m - 1$, for all j such that*

$$0 < W_q(j) \le (q-1)m - r - 1.$$

The code of block length q^m obtained by the expansion of this cyclic code by an overall parity-check is called a GRM code of order r and block length q^m.

EXAMPLE 11.21 *Let $q = 2, m = 3$, and $r = 1$. Then $n = 2^3 - 1 = 7$ and*

$$0 < W_2(j) \le m - r - 1 = 3 - 1 - 1 = 1,$$

i.e., $j = 1, 2, 4$. Let α be a primitive element in $\mathrm{GF}(8)$ such that $\alpha^3 = \alpha + 1$. It follows that

$$g(x) = (x - \alpha)(x - \alpha^2)(x - \alpha^4) = x^3 + x + 1.$$

For this code $d = 2^{m-r} - 1 = 2^{3-1} - 1 = 3$.

THEOREM 11.22 *The cyclic binary GRM code of order r and block length $n = 2^m - 1$ is a subcode of the primitive BCH code of design distance $d = 2^{m-r} - 1$.*

Proof: The number $2^{m-r} - 1$ is represented in radix-2 by an $(m-r)$-bit number consisting of $m - r$ ones. Any positive integer j smaller than $2^{m-r} - 1$ will have less than $m - r$ ones in the respective radix-2 representation. It thus follows that

$$W_2(j) \le m - r - 1, \text{ for } j = 1, 2, \ldots, 2^{m-r} - 2,$$

and thus α^j will be a root of $g(x)$ for those j. The code is therefore a subcode of the primitive BCH code of design distance $2^{m-r} - 1$ because the α^j, $j = 1, 2, \ldots, 2^{m-r} - 2$ consecutive roots guarantee that the generator polynomial of the GRM code of order r is divisible by the generator polynomial for the primitive BCH code with design distance $2^{m-r} - 1$.   □

The minimum distance of these binary GRM codes is at least $2^{m-r} - 1$, as follows from Theorem 11.23, since they are subcodes of BCH codes of design distance $2^{m-r} - 1$.

THEOREM 11.23 *The dual of a GRM code over $\mathrm{GF}(q)$, of order r and block length q^m, is the GRM code over $\mathrm{GF}(q)$ of order $r' = (q-1)m - r - 1$ and block length q^m.*

Proof: Let C_{er} denote the GRM code of order r and let C_r denote the cyclic GRM code obtained by shortening C_{er}. The proof begins by finding $g^\perp(x)$, the generator polynomial for the code $C_r^\perp$, dual of C_r. Let α be a primitive element of $\mathrm{GF}(q^m)$. The generator polynomial of C_r has roots α^j, $1 \le j \le q^m - 1$, for all j satisfying $0 < W_q(j) \le (q-1)m - r - 1$. Notice that the roots of the parity-check polynomial $h^\perp(x)$, of $C_r^\perp$, are powers of α obtained by replacing j in α^j by $q^m - 1 - j$ satisfying the inequality $W_q(q^m - 1 - j) > (q - 1)m - r - 1$. This is the case because $h^\perp(x)$ is given by the reciprocal of the generator polynomial $g(x)$ of code C_r. It is now observed that if $j + j' = q^m - 1$ then

$$W_q(j + j') = W_q(j) + W_q(j') = W_q(q^m - 1) = (q - 1)m$$

or

$$W_q(j) = (q - 1)m - W_q(q^m - 1 - j),$$

and combining this result with the inequality

$$W_q(q^m - 1 - j) > (q - 1)m - r - 1$$

it follows that

$$W_q(j) < r + 1$$

which characterizes j in α^j, the roots of $g^\perp(x)$, including $\alpha^0 = 1$ as a root of $g^\perp(x)$. The cyclic GRM code of block length $q^m - 1$ and order $r' = (q - 1)m - r - 1$, denoted as $C_{r'}$, has as roots of its generator polynomial α^j, $1 \le j \le q^m - 1$, where j satisfies

$$0 < W_q(j) \le (q - 1)m - (q - 1)m + r + 1 - 1 = r.$$

It is noticed that, with the exception of $\alpha^0 = 1$, the roots of the generator polynomial of code $C_{r'}$ are the same as those of $g^\perp(x)$ found earlier, i.e., $C_{r'}$ contains $C_r^\perp$. It is further observed that both codes C_r and $C_{r'}$ have the all ones codeword since $\alpha^0 = 1$ is not a root in their respective generator polynomials. The generator matrices for these codes are considered in the following form:

$$G = \underbrace{\left[\begin{array}{cccc} 1 & 1 & \cdots & 1 \\ & G_1 & & \end{array} \right]}_{q^m - 1 \text{ columns}}$$

where G_1 denotes the generator matrix for the subcode containing the root $\alpha^0 = 1$ in its generator polynomial. By extending the generator matrices for codes C_r and C'_r the result is

$$
G_{er} =
\begin{bmatrix}
1 & 1 & \cdots & 1 & \vdots & 1 \\
 & & & & \vdots & 0 \\
 & G_{1r} & & & \vdots & \vdots \\
 & & & & \vdots & 0
\end{bmatrix}
\qquad
G_{er'} =
\begin{bmatrix}
1 & 1 & \cdots & 1 & \vdots & 1 \\
 & & & & \vdots & 0 \\
 & G_{1r'} & & & \vdots & \vdots \\
 & & & & \vdots & 0
\end{bmatrix}
$$

where the row space of G_{er} is orthogonal to the row space of $G_{er'}$ and the sum of the code dimensions is q^m, therefore codes G_{er} and $G_{er'}$ are dual.   □

The reader may wish to prove Theorem 11.23 by assuming that the row space of G_{er} is orthogonal to the row space of $G_{er'}$.

11.9   Euclidean geometry codes

DEFINITION 11.24 *Let* r, *s and* m *be positive integers, and let* $q = p^s$, *where* p *is a prime number. The Euclidean geometry code over* GF(p), *of block length* $n = q^m$ *and order* r, *is the dual of the subfield subcode of the GRM code over* GF(q) *of block length* q^m *and order* $(q - 1)(m - r - 1)$.

Equivalently, a Euclidean geometry code can be characterized by the following theorem.

THEOREM 11.25 *Let* α *be a primitive element of* GF(p^{sm}). *A Euclidean geometry code over* GF(p), *with parameters* r, s, *and block length* q^m *is obtained by extending a cyclic code the generator polynomial of which has* α^j *for a root,* $1 \le j \le q^m - 1$, *if* j *satisfies the inequality*

$$
0 < \max_{0 \le i < s} W_q(jp^i) \le (q - 1)(m - r - 1).
$$

Proof: The roots $\alpha^j, 1 \le j \le q^m - 1$, of the generator polynomial of the GF(p) subfield subcode of the cyclic GRM code of order $(q-1)(m-r-1)$ are characterized by those values of j and its p-ary conjugates satisfying

$$
1 \le \max_{0 \le i < s} W_q(jp^i) \le (q - 1)m - (q - 1)(m - r - 1) - 1
$$

$$
= (q - 1)(r + 1) - 1.
$$

Conversely, α^j is a root of the parity-check polynomial $h(x)$ if

$$
\max_{0 \le i < s} W_q(jp^i) > (q - 1)(r + 1) - 1,
$$

or if $j = 0$. It is well known that the generator polynomial $g^\perp(x)$ of the dual code is the reciprocal polynomial of $h(x)$. Then $\alpha^{j'}$ is a root of $g^\perp(x)$ if $\alpha^{n-j'}$ is a root of $h(x)$, that is, if

$$W_q(n - j') > (q - 1)(r + 1) - 1$$

or if $n - j' = 0$, i.e., if $j' = 0 \mod n$ for every j' that is a p-ary conjugate of j. However, to obtain the correct extended code it is necessary to consider a cyclic code whose generator polynomial has the same roots as $g^\perp(x)$, except for the root equal to 1, i.e., the value $j' = 0$ is disregarded. It is noticed in the proof of Theorem 11.23 that

$$W_q(n - j') + W_q(j') = W_q(q^m - 1) = (q - 1)m.$$

Hence, it follows that

$$W_q(j') < (q - 1)m - (q - 1)(r + 1) + 1 = (q - 1)(m - r - 1) + 1$$

or

$$1 \le W_q(j') \le (q - 1)(m - r - 1)$$

for every j' that is a p-ary conjugate of j. $\qquad\square$

In order to derive some important properties of Euclidean geometry (EG) codes, e.g., its threshold decoding, the value of J and the number of decoding steps, some concepts from Euclidean geometry are required.

A Euclidean geometry $EG(m, q)$ of dimension m over $GF(q)$, consists of the set of q^m points from the vector space $GF(q)^m$, together with all its subspaces and translation of subspaces called *flats*, as defined in the sequel.

Concepts from Euclidean geometry

DEFINITION 11.26 Affine subspace
An affine subspace is a translation of a vector subspace, i.e., a coset.

DEFINITION 11.27 t-flat
Let $v_0, v_1, \ldots, v_t$ be $t + 1$ linearly independent points in $EG(m, q)$, where $m > t$. The set of q^t points defined by

$$v_0 + \beta_1 v_1 + \beta_2 v_2 + \cdots + \beta_t v_t \tag{11.12}$$

with $\beta_i \in GF(q), 1 \le i \le t$, constitute a t-flat, or a t-dimensional hyperplane, which passes through the point v_0.

$(t + 1)$-flats can be obtained recursively as distinct translations of the smallest vector subspace containing a t-flat. That is, if E_t is a t-flat then the set $\{\beta_t v | \beta_t \in \mathrm{GF}(q), v \in E_t\}$ is the smallest subspace containing E_t. If $v_{t+1} \notin E_t$ is a point in $\mathrm{GF}(q)^m$, then a $(t+1)$-flat E_{t+1} is obtained as the set

$$E_{t+1} = \{v_{t+1} + \beta_t v | \beta_t \in \mathrm{GF}(q), v \in E_t\}.$$

DEFINITION 11.28 **Gaussian coefficients**
The q-ary Gaussian coefficients are defined as

$$\begin{bmatrix} m \\ i \end{bmatrix} = \prod_{j=0}^{i-1} \frac{q^m - q^j}{q^i - q^j}, \quad i = 0, 1, \ldots, m, \text{ and} \quad \begin{bmatrix} m \\ 0 \end{bmatrix} = 1.$$

THEOREM 11.29 *$EG(m, q)$ contains $q^{m-t} \begin{bmatrix} m \\ t \end{bmatrix}$ distinct t-flats, for $t = 1, 2, \ldots, m - 1$.*

Proof: A t-dimensional subspace can be constructed by choosing a set of t linearly independent points in $EG(m, q)$. The number of different ways of choosing such t points is given by

$$(q^m - 1)(q^m - q) \ldots (q^m - q^{t-1}). \tag{11.13}$$

However, a number of such sets of t points generate the same t-dimensional subspace. The number of identical t-dimensional subspaces is given by

$$(q^t - 1)(q^t - q) \ldots (q^t - q^{t-1}). \tag{11.14}$$

The ratio between (11.13) and (11.14) is represented by $\begin{bmatrix} m \\ t \end{bmatrix}$, and gives the number of distinct t-dimensional subspaces in $EG(m, q)$. Each subspace has q^{m-t} cosets, i.e., possible values of v_0 in (11.12), and thus there are $q^{m-t} \begin{bmatrix} m \\ t \end{bmatrix}$ t-flats in $EG(m, q)$. $\qquad \square$

THEOREM 11.30 *For any integers s and t, $0 \leq s \leq t < m$, each s-flat is contained in $\begin{bmatrix} m - s \\ t - s \end{bmatrix}$ distinct t-flats in $EG(m, q)$.*

Proof: Given an s-flat, it can be extended to a t-flat by choosing $t - s$ linearly independent points from $\mathrm{GF}(q)^m$, not yet included in the s-flat. The number of possibilities for doing this is given by

$$(q^m - q^s)(q^m - q^{s+1}) \ldots (q^m - q^{t-1}). \tag{11.15}$$

However, some sequences of such $t - s$ linearly independent points will generate the same t-flat, precisely

$$(q^t - q^s)(q^t - q^{s+1}) \ldots (q^t - q^{t-1}), \tag{11.16}$$

where the $t - s$ points are chosen not using points from the s-flat. The ratio between (11.15) and (11.16), denoted as $\begin{bmatrix} m - s \\ t - s \end{bmatrix}$ gives the desired result. □

DEFINITION 11.31 *Given a subset of elements of a set of size n, an n-coordinate binary vector is constructed containing a "1" in the ith coordinate if i is in the subset, otherwise a zero is placed in the ith coordinate. Such a vector is called the incidence vector of the subset.*

LEMMA 11.32 *Given two non-negative integers n and k, and a prime p, if k is a radix-p descendent of n then k is also a radix-q descendent of n for $q = p^s$, where $s > 0$ is an integer.*

THEOREM 11.33 *An EG code of order r and length q^m, over $\mathrm{GF}(p)$, is the largest linear code over $\mathrm{GF}(p)$ having in its null space the incidence vectors of all $(r + 1)$-flats in $\mathrm{EG}(m, q)$.*

Proof: The proof consists of showing that the GRM code of order $(q - 1)(m - r - 1)$, containing the dual of the EG code, is the smallest linear code over $\mathrm{GF}(q)$ that contains all the incidence vectors stated in the theorem. An incidence vector is in the GRM code if

(a) It is in the associated cyclic GRM code

(b) It has the correct extension symbol.

Notice that the incidence vector of an $(r + 1)$-flat contains q^{r+1} nonzero components, all of which are 1's, and thus add to zero modulo p, since $q = p^s$. The extension symbol is thus always correct. Now it is only necessary to prove that the incidence vector, with the extension symbol deleted, belongs to the cyclic GRM code. This last vector will be referred to as the *shortened incidence vector*. One way of proving that is by showing that the shortened incidence vectors have in common the roots $\alpha^j, 1 \leq j \leq q^m - 1$, where j satisfies

$$1 \leq W_q(j) \leq (q - 1)m - (q - 1)(m - r - 1) - 1 = (q - 1)(r + 1) - 1,$$

and that, for other values of j, α^j is not among their common roots. Consider now the $(r + 1)$-flat

$$v_0 + \beta_1 v_1 + \beta_2 v_2 + \cdots + \beta_{r+1} v_{r+1},$$

where $\beta_i \in \mathrm{GF}(q)$, $1 \leq i \leq r + 1$, and $v_0, v_1, \ldots, v_{r+1}$ are a set of $r + 2$ linearly independent points in the $(r + 1)$-flat.

We denote a shortened incidence vector by a polynomial in x, $f(x)$, as follows:

$$f(x) = \sum_{i=0}^{n-1} f_i x^i,$$

where $f_i, 0 \le i \le n-1$, are the coordinates of the shortened incidence vector, i.e., f_i is either equal to zero or equal to one. If α^j is a root of $f(x)$, it follows that

$$f(\alpha^j) = \sum_{i=0}^{n-1} f_i \alpha^{ij} = 0$$

or, in terms of the $(r+1)$-flat,

$$f(\alpha^j) = \sum_{\beta_1=0}^{q-1} \sum_{\beta_2=0}^{q-1} \cdots \sum_{\beta_{r+1}=0}^{q-1} (v_0 + \beta_1 v_1 + \beta_2 v_2 + \cdots + \beta_{r+1} v_{r+1})^j = 0$$

since $\alpha^i = v_0 + \beta_1 v_1 + \beta_2 v_2 + \cdots + \beta_{r+1} v_{r+1}$, if $f_i = 1$.

The next step is to determine those j for which $f(\alpha^j) = 0$. A multinomial expansion of $f(\alpha^j)$ gives

$$f(\alpha^j) = \sum_{\beta_1=0}^{q-1} \cdots \sum_{\beta_{r+1}=0}^{q-1} \sum_h \frac{j!}{h_0! h_1! \ldots h_{r+1}!} v_0^{h_0} (\beta_1 v_1)^{h_1} \ldots$$
$$\ldots (\beta_{r+1} v_{r+1})^{h_{r+1}},$$

where the summation on h is over all $(r+2)$-tuples $(h_0, h_1, \ldots, h_{r+1})$ such that $\sum_{i=0}^{r+1} h_i = j$. Interchanging the order of summation, to work with terms of the form $\sum_{\beta_i=0}^{q-1} (\beta_i v)^h$ with h fixed, the following lemma is used next.

LEMMA 11.34

If $h \ne 0$ then $\sum_{\beta_i=0}^{q-1} (\beta_i)^h = \begin{cases} -1, & \text{if } h = a(q-1), a \ne 0 \\ 0, & \text{otherwise} \end{cases}$

and

if $h = 0$ then $\sum_{\beta_i=0}^{q-1} (\beta_i)^h = 0$.

Therefore, every term in the sum for $f(\alpha^j)$ is zero except when h_l is a nonzero multiple of $(q-1)$, for $l = 1, 2, \ldots, r+1$. The sum for $f(\alpha^j)$ can thus be written as

$$f(\alpha^j) = (-1)^{r+1} \sum_{h_1} \sum_{h_2} \cdots \sum_{h_{r+1}} \frac{j!}{h_0! h_1! \ldots h_{r+1}!} v_0^{h_0} v_1^{h_1} \ldots v_{r+1}^{h_{r+1}},$$

where the sum is over all $(h_0, h_1, \ldots, h_{r+1})$ such that $\sum_{l=0}^{r+1} h_j = j$, h_l is a nonzero multiple of $q - 1$ for $l = 1, 2, \ldots, r + 1$, and $h_0 \geq 0$. Let the multinomial coefficients be expressed as follows:

$$\frac{j!}{h_0! h_1! \ldots h_{r+1}!} = \frac{j!}{h_0!(j - h_0)! \, h_1! \ldots h_{r+1}!} \frac{(j - h_0)!}{}$$

$$= \frac{j!}{h_0!(j - h_0)!} \frac{(j - h_0)!}{h_1!(j - h_0 - h_1)!} \frac{(1 - h_0 - h_1)!}{h_2! \ldots h_{r+1}!} = \text{etc.}$$

By Lucas's theorem, if h_l contributes to the sum then h_l is a radix-p descendent of j for $l = 0, 1, \ldots, r + 1$, and it follows that h_l is also a radix-q descendent of j for $l = 0, 1, \ldots, r + 1$. Furthermore, any sum of h_j's, contributing to the sum for $f(\alpha^j)$, is a radix-p descendent of j and therefore is a radix-q descendent of j, for $q = p^s$.

Summary of conditions on terms that contribute to the sum for $f(\alpha^j)$:

(1) $\sum_{l=0}^{r+1} h_j = j$

(2) $h_l = a(q - 1), a > 0, l \in \{1, 2, \ldots, r + 1\}, h_0 \geq 0.$

(3) Each coefficient of q^i, $0 \leq i \leq m - 1$, in the radix-q expansion of j is the sum of the corresponding coefficients of the radix-q expansion of the h_l's, and it follows that

$$W_q(j) = \sum_{l=0}^{r+1} W_q(h_l). \tag{11.17}$$

Finally, it is shown next that there are no terms that contribute to the sum for $f(\alpha^j)$ if $W_q(j) < (q - 1)(r + 1)$.

LEMMA 11.35 *Consider the radix-q expansion of some integer k, i.e.,*

$$k = k_0 + k_1 q + k_2 q^2 + \cdots + k_{m-1} q^{m-1}$$

and the radix-q weight of k, i.e.,

$$W_q(k) = k_0 + k_1 + k_2 + \cdots + k_{m-1}.$$

Then k is a nonzero multiple of $q - 1$ if and only if $W_q(k)$ is also a nonzero multiple of $q - 1$.

Proof: Consider the difference $k - W_q(k)$, i.e.,

$$k - W_q(k) = k_1(q - 1) + k_2(q^2 - 1) + \cdots + k_{m-1}(q^{m-1} - 1) \tag{11.18}$$

The lemma follows by noticing that if either of the terms on the left side in (11.18) is divisible by $q-1$ then the other is also divisible by $q-1$.

□

However, it follows from (11.17) and the fact that $W_q(h_l)$ is a nonzero multiple of $q-1$, for $l=1,2,\ldots,r+1$, that

$$W_q(j) = \sum_{l=0}^{r+1} W_q(h_l) \geq (q-1)(r+1),$$

where the j's satisfying this inequality cause $f(\alpha^j) \neq 0$. The theorem is thus proved.

□

THEOREM 11.36 *Let* $q = p^s$. *The EG code of order* r *and block length* q^m *over* $\mathrm{GF}(p)$, *can be* L-*step threshold decoded in* $L = r+1$ *steps, with ability to correct at least* $\left\lfloor \frac{1}{2}\frac{(q^{m-r}-1)}{(q-1)} \right\rfloor$ *errors.*

Proof: Consider the incidence vectors of the $(r+1)$-flats, i.e., a set of codewords in the dual code. The incidence vectors of all $(r+1)$-flats that contain a given r-flat define a set of

$$\begin{bmatrix} m-r \\ r+1-r \end{bmatrix} = \begin{bmatrix} m-r \\ 1 \end{bmatrix} = \frac{q^{m-r}-1}{q-1}$$

parity-check sums orthogonal on a sum of error digits associated with the points of the r-flat. Using these $(q^{m-r}-1)/(q-1)$ parity-check sums an estimate of the sum of error digits is obtained by majority voting, where it is noticed that, for a given r-flat E, any point not belonging to E is contained in a distinct $(r+1)$-flat that contains E. In this manner, a new parity-check sum is obtained which corresponds to the incidence vector of the r-flat E. This procedure can be repeated for all r-flats that contain a given $(r-1)$-flat, which in turn define a set of parity-check sums orthogonal on that $(r-1)$-flat. By induction, after $r+1$ steps, a set of parity-check sums is obtained which is orthogonal on a single error digit (associated with a 0-flat). The number of parity-check sums orthogonal on a given sum of error digits is given at each decoding step by

$$\underbrace{\begin{bmatrix} m-r \\ 1 \end{bmatrix}}_{\text{step }1} < \underbrace{\begin{bmatrix} m-(r-1) \\ 1 \end{bmatrix}}_{\text{step }2} < \underbrace{\begin{bmatrix} m-(r-2) \\ 1 \end{bmatrix}}_{\text{step }3} \cdots < \underbrace{\begin{bmatrix} m-(r-r) \\ 1 \end{bmatrix}}_{\text{step }r+1}$$

and therefore the number of errors that can be corrected with this procedure is at least $\left\lfloor \frac{1}{2}\frac{(q^{m-r}-1)}{(q-1)} \right\rfloor$.

□

11.10   Projective geometry codes

These codes are constructed from nonprimitive GRM codes of length $(q^m - 1)/(q - 1)$.

DEFINITION 11.37 *Let r, s and m be any positive integers, and let $q = p^s$, where p is a prime. The projective geometry (PG) code over $\mathrm{GF}(p)$ of length $n = (q^m - 1)/(q - 1)$ and order r is the dual of the subfield subcode of the nonprimitive cyclic GRM code over $\mathrm{GF}(q)$ of the same block length and order $(q - 1)(m - r - 1)$.*

DEFINITION 11.38 (**Nonprimitive GRM codes**) *Let b be a nontrivial factor of $q^m - 1$. The cyclic nonprimitive GRM code of order r and block length $(q^m - 1)/b$ over $\mathrm{GF}(q)$ is the cyclic code whose generator polynomial has zeros at α^{jb}, $j = 1, 2, \ldots, (q^m - 1)/b$, such that $0 < W_q(b_j) \leq (q - 1)m - r - 1$.*

The generator polynomial of a PG code is characterized by the following theorem.

THEOREM 11.39 *A PG code over $\mathrm{GF}(p)$ of block length $(q^m - 1)/(q - 1)$ and order r, with $q = p^s$, is the cyclic code whose generator polynomial has zeros at β^j, $0 < j \leq (q^m - 1)/(q - 1)$, if j satisfies*

$$0 < W_q[j(q - 1)p^i] \leq (q - 1)(m - r - 1)$$

where $\beta = \alpha^{q-1}$, and α is primitive in $\mathrm{GF}(q^m)$.

PG codes can be threshold decoded by using concepts from projective geometry. The projective geometry $\mathrm{PG}(m, q)$ has $(q^m - 1)/(q - 1)$ points and is defined by using the nonzero elements of $\mathrm{GF}(q)^{m+1}$. The $q^{m+1} - 1$ nonzero elements of $\mathrm{GF}(q)^{m+1}$ are divided into $(q^m - 1)/(q - 1)$ sets, each set represents one point of $\mathrm{PG}(m, q)$ and can be expressed as $\{\lambda \mathbf{v} | \mathbf{v} \neq \mathbf{0}, \mathbf{v} \in \mathrm{GF}(q)^{m+1}, \lambda \in \mathrm{GF}(q), \lambda \neq 0\}$. The $\mathrm{PG}(m, q)$ of dimension m over $\mathrm{GF}(q)$ is the set of these $(q^m - 1)/(q - 1)$ points together with subsets of points called t-flats, for $t = 0, 1, 2, \ldots, m$.

Let $\mathbf{v}_i \in V_i$, for $i = 0, 1, \ldots, t$. Consider $t + 1$ linearly independent points $\mathbf{v}_i$ in $\mathrm{GF}(q)^{m+1}$. A t-flat containing $V_0, V_1, \ldots, V_t$ is the set of points in $\mathrm{PG}(m, q)$ which are the images of the points

$$\beta_0 \mathbf{v}_0 + \beta_1 \mathbf{v}_1 + \cdots + \beta_t \mathbf{v}_t,$$

in $\mathrm{GF}(q)^{m+1}$, where $\beta_i, 0 \leq i \leq t$, are arbitrary $\mathrm{GF}(q)$ elements not all zero.

THEOREM 11.40 *The flats in a projective geometry satisfy the following:*

(i) $PG(m, q)$ *contains* $\begin{bmatrix} m + t \\ t + 1 \end{bmatrix}$ *distinct t-flats,* $t = 0, 1, \ldots, m.$

(ii) *For any* s *and* t, $0 \leq s \leq t \leq m$, *each s-flat is contained in* $\begin{bmatrix} m - s \\ t - s \end{bmatrix}$ *distinct t-flats in* $PG(m, q)$.

THEOREM 11.41 *A PG code of order* r *and block length* $(q^m - 1)/(q - 1)$ *over* $GF(p)$ *is the largest linear code over* $GF(p)$ *having in its null space the incidence vectors of all r-flats in* $PG(m, q)$.

Projective geometry codes can be L-step threshold decoded and can correct at least $\lfloor \frac{1}{2}(q^{m-r-1} - 1)/(q - 1) \rfloor$ errors.

11.11   Problems with solutions

(1) Consider the $(7, 3, 4)$ binary cyclic code with generator polynomial $g(x) = x^4 + x^2 + x + 1$, which is the dual of the Hamming $(7, 4, 3)$ code. Derive the parity-check matrix for this $(7, 3, 4)$ code in reduced echelon form.

Solution: The parity-check matrix for this $(7, 3, 4)$ code in reduced echelon form, is the following:

$$\mathbf{H} = \begin{bmatrix} 1 & 1 & 0 & 1 & 0 & 0 & 0 \\ 0 & 1 & 1 & 0 & 1 & 0 & 0 \\ 1 & 1 & 1 & 0 & 0 & 1 & 0 \\ 1 & 0 & 1 & 0 & 0 & 0 & 1 \end{bmatrix}.$$

(2) Derive a set of three parity-sums orthogonal on position e_6 for the code of Problem 1.

Solution:
$$\begin{array}{rcll} A_1 & = & e_6 + e_5 + e_3 & \text{row 1} \\ A_2 & = & e_6 + e_2 + e_1 & \text{rows } 2 \oplus 3 \\ A_3 & = & e_6 + e_4 + e_0 & \text{row 4} \end{array}$$

(3) Let $m = 3$ and for $n = 2^m = 8$ construct the matrices $\mathbf{G}_0$, $\mathbf{G}_1$, and $\mathbf{G}_2$, which are building blocks for RM codes of block length 8.

Solution:
$$\mathbf{G}_0 = \begin{bmatrix} 1 & 1 & 1 & 1 & 1 & 1 & 1 & 1 \end{bmatrix}$$

$$G_1 = \begin{bmatrix} 0 & 0 & 0 & 0 & 1 & 1 & 1 & 1 \\ 0 & 0 & 1 & 1 & 0 & 0 & 1 & 1 \\ 0 & 1 & 0 & 1 & 0 & 1 & 0 & 1 \end{bmatrix}$$

$$G_2 = \begin{bmatrix} 0 & 0 & 0 & 0 & 0 & 0 & 1 & 1 \\ 0 & 0 & 0 & 0 & 0 & 1 & 0 & 1 \\ 0 & 0 & 0 & 1 & 0 & 0 & 0 & 1 \end{bmatrix}$$

(4) Write the generator matrix for the first-order RM code of block length 8.

Solution: The generator matrix for the first-order RM code of block length 8 is the following 4×8 matrix:

$$G = \begin{bmatrix} \mathbf{G_0} \\ \mathbf{G_1} \end{bmatrix} = \begin{bmatrix} 1 & 1 & 1 & 1 & 1 & 1 & 1 & 1 \\ 0 & 0 & 0 & 0 & 1 & 1 & 1 & 1 \\ 0 & 0 & 1 & 1 & 0 & 0 & 1 & 1 \\ 0 & 1 & 0 & 1 & 0 & 1 & 0 & 1 \end{bmatrix} \begin{matrix} k_1 \\ k_2 \\ k_3 \\ k_4 \end{matrix}$$

$$0 \quad 1 \quad 2 \quad 3 \quad 4 \quad 5 \quad 6 \quad 7$$

which generates the $(7,4,3)$ Hamming code extended by an overall parity-check bit to $(8,4,4)$.

(5) Decode using the Reed algorithm (by majority voting) the $\mathbf{G_1}$ segment for the code of Problem 4.

Solution: Decoding by the Reed algorithm

(a) Decoding the $\mathbf{G_1}$ segment by majority voting, denoted as MAJ. Suppose the received n-tuple is $\mathbf{z} = (z_0, z_1, z_2, z_3, z_4, z_5, z_6, z_7)$.

$$k_4 : \quad \mathrm{MAJ}(z_0 \oplus z_1, z_2 \oplus z_3, z_4 \oplus z_5, z_6 \oplus z_7)$$
$$k_3 : \quad \mathrm{MAJ}(z_0 \oplus z_2, z_1 \oplus z_3, z_4 \oplus z_6, z_5 \oplus z_7)$$
$$k_2 : \quad \mathrm{MAJ}(z_0 \oplus z_4, z_1 \oplus z_5, z_2 \oplus z_6, z_3 \oplus z_7)$$

(b) Remove the effect of k_4, k_3, and k_2 from the received n-tuple $\mathbf{z}$. The result is an n-tuple $\mathbf{z'}$ from the $r-1 = 1-1 = 0$th order RM code of block length 8, which is decoded by taking a majority vote among its coordinates, i.e.,

$$k_1 = \mathrm{MAJ}(z_0', z_1', z_2', z_3', z_4', z_5', z_6', z_7').$$

(6) Let $m = 2$ and $s = 2$ and consider the Euclidean geometry $EG(m, 2^s)$ over $GF(2^s)$. Determine the number of points and the number of lines in $EG(2, 4)$ over $GF(4)$.

Table 11.1.   Points in EG$(2, 4)$ over GF(4).

$\mathbf{x}_0 = (0, 0),$	$\mathbf{x}_1 = (0, 1),$	$\mathbf{x}_2 = (0, \alpha),$	$\mathbf{x}_3 = (0, \alpha^2),$
$\mathbf{x}_4 = (1, 0),$	$\mathbf{x}_5 = (1, 1),$	$\mathbf{x}_6 = (1, \alpha),$	$\mathbf{x}_7 = (1, \alpha^2),$
$\mathbf{x}_8 = (\alpha, 0),$	$\mathbf{x}_9 = (\alpha, 1),$	$\mathbf{x}_{10} = (\alpha, \alpha),$	$\mathbf{x}_{11} = (\alpha, \alpha^2),$
$\mathbf{x}_{12} = (\alpha^2, 0),$	$\mathbf{x}_{13} = (\alpha^2, 1),$	$\mathbf{x}_{14} = (\alpha^2, \alpha),$	$\mathbf{x}_{15} = (\alpha^2, \alpha^2)$

Table 11.2.   Lines in EG$(2, 4)$ over GF(4).

$\{\mathbf{x}_0, \mathbf{x}_1, \mathbf{x}_2, \mathbf{x}_3\}$	$\{\mathbf{x}_4, \mathbf{x}_5, \mathbf{x}_6, \mathbf{x}_7\}$	$\{\mathbf{x}_8, \mathbf{x}_9, \mathbf{x}_{10}, \mathbf{x}_{11}\}$	$\{\mathbf{x}_{12}, \mathbf{x}_{13}, \mathbf{x}_{14}, \mathbf{x}_{15}\}$
$\{\mathbf{x}_0, \mathbf{x}_4, \mathbf{x}_8, \mathbf{x}_{12}\}$	$\{\mathbf{x}_1, \mathbf{x}_5, \mathbf{x}_9, \mathbf{x}_{13}\}$	$\{\mathbf{x}_2, \mathbf{x}_6, \mathbf{x}_{10}, \mathbf{x}_{14}\}$	$\{\mathbf{x}_3, \mathbf{x}_7, \mathbf{x}_{11}, \mathbf{x}_{15}\}$
$\{\mathbf{x}_0, \mathbf{x}_5, \mathbf{x}_{10}, \mathbf{x}_{15}\}$	$\{\mathbf{x}_1, \mathbf{x}_4, \mathbf{x}_{11}, \mathbf{x}_{14}\}$	$\{\mathbf{x}_2, \mathbf{x}_7, \mathbf{x}_8, \mathbf{x}_{13}\}$	$\{\mathbf{x}_3, \mathbf{x}_6, \mathbf{x}_9, \mathbf{x}_{12}\}$
$\{\mathbf{x}_0, \mathbf{x}_6, \mathbf{x}_{11}, \mathbf{x}_{13}\}$	$\{\mathbf{x}_1, \mathbf{x}_7, \mathbf{x}_{10}, \mathbf{x}_{12}\}$	$\{\mathbf{x}_2, \mathbf{x}_4, \mathbf{x}_9, \mathbf{x}_{15}\}$	$\{\mathbf{x}_3, \mathbf{x}_5, \mathbf{x}_8, \mathbf{x}_{14}\}$
$\{\mathbf{x}_0, \mathbf{x}_7, \mathbf{x}_9, \mathbf{x}_{14}\}$	$\{\mathbf{x}_1, \mathbf{x}_6, \mathbf{x}_8, \mathbf{x}_{15}\}$	$\{\mathbf{x}_2, \mathbf{x}_5, \mathbf{x}_{11}, \mathbf{x}_{12}\}$	$\{\mathbf{x}_3, \mathbf{x}_4, \mathbf{x}_{10}, \mathbf{x}_{13}\}$

Solution: The number of points in EG$(m, 2^s)$ over GF(2^s) is 2^{ms}, i.e., $(2^2)^2 = 16$ points, indicated in Table 11.1. The corresponding number of lines (1-flat, see (11.12)) is $2^{(m-1)s}(2^{ms}-1)/(2^s-1) = 20$, and are indicated in Table 11.2. Let α be a primitive element of GF(4) whose minimal polynomial is $x^2 + x + 1$, i.e. $\alpha^2 = \alpha + 1$. The elements of GF(4) are denoted as $\{0, 1, \alpha, \alpha^2\}$.

Applying the construction rule indicated in the proof of Theorem 11.29, we obtain $(2^{ms} - 1)/(2^s - 1) = 5$ linearly independent points, namely, $\mathbf{x}_1, \mathbf{x}_4, \mathbf{x}_5, \mathbf{x}_6$, and $\mathbf{x}_7$. We obtain the following five subspaces, $\{\mathbf{x}_0, \mathbf{x}_1, \mathbf{x}_2, \mathbf{x}_3\}$, $\{\mathbf{x}_0, \mathbf{x}_4, \mathbf{x}_8, \mathbf{x}_{12}\}$, $\{\mathbf{x}_0, \mathbf{x}_5, \mathbf{x}_{10}, \mathbf{x}_{15}\}$, $\{\mathbf{x}_0, \mathbf{x}_6, \mathbf{x}_{11}, \mathbf{x}_{13}\}$, and $\{\mathbf{x}_0, \mathbf{x}_7, \mathbf{x}_9, \mathbf{x}_{14}\}$.

Appendix A
The Gilbert Bound

A.1   Introduction

Let A be any finite set and let A^n represent the set of all n-tuples $\mathbf{v} = (v_1, v_2, \ldots, v_n)$ with components in A. Let $|A| = q$.

The Hamming sphere of radius t, non-negative, with center at the n-tuple $\mathbf{v} \in A^n$, is the set of all vectors $\mathbf{v}^*$, $\mathbf{v}^* \in A^n$, such that $d_H(\mathbf{v}, \mathbf{v}^*) \leq t$, where $d_H(\mathbf{v}, \mathbf{v}^*)$ is the Hamming distance between $\mathbf{v}$ and $\mathbf{v}^*$.

The number of points, S_t, in a Hamming sphere of radius t is given by

$$S_t = \sum_{i=0}^{t} \binom{n}{i}(q-1)^i.$$

For given positive integers n and d, $1 \leq d \leq n$, a block code with minimum Hamming distance $d_{min} \geq d$ can be constructed as follows. Choose one n-tuple $\mathbf{v}_1 \in A^n$, arbitrarily among the q^n n-tuples in A^n. Discard all n-tuples $\mathbf{v}^* \in A^n$ which are in the Hamming sphere of radius $d-1$ centered at $\mathbf{v}_1$. Repeat the procedure with another n-tuple $\mathbf{v}_2, \mathbf{v}_2 \neq \mathbf{v}_1$, selected from among the "surviving" n-tuples. This procedure should be carried out as long as possible, at each step eliminating at most S_{d-1} n-tuples. The final subset of n-tuples, say of size M, which constitutes the code, must satisfy

$$S_{d-1}M \geq q^n,$$

or

$$S_{d-1} = \sum_{i=0}^{d-1} \binom{n}{i}(q-1)^i \geq q^{n(1-R)}, \tag{A.1}$$

where R is the code rate and $M = q^{nR}$. Expression (A.1) is known as the Gilbert lower bound on d, which can be stated as follows.

Gilbert lower bound: For any positive integers n and d, $1 \leq d \leq n$, there exists a q-ary block code with $d_{min} \geq d$, whose rate R satisfies $\sum_{i=0}^{d-1} \binom{n}{i}(q-1)^i \geq q^{n(1-R)}$.

A.2   The binary asymptotic Gilbert bound

In practice, the greatest interest focus on the case where $q = 2$, i.e., the binary case. The following inequalities (Massey 1985) hold for S_t

$$(t/\sqrt{2n})2^{nh(t/n)} < S_t \leq 2^{nh(t/n)}, \text{ if } t/n \leq 1/2,$$

where $h(p) = -p \log_2 p - (1-p)\log_2(1-p)$, $0 < p < 1$, is the binary entropy function (Cover and Thomas 2006, p.14). Thus, with the help of (A.1) we have

$$2^{n(1-R)} < S_{d-1} \leq 2^{nh[(d-1)/n]}, \text{ or } 1 - R \leq h[(d-1)/n].$$

When n is sufficiently large the rate R satisfies

$$R \geq 1 - h(d/n), \text{ if } d_{min}/n \leq 1/2.$$

This is the asymptotic Gilbert bound, which holds for all n. As noted in (Massey 1985), the binary asymptotic Gilbert bound has now resisted for over 50 years all attempts to improve it. Curiously enough, it can be shown that with probability approaching 1 as n approaches infinite, a binary code of rate R obtained by the statistically independent selection of equally likely codewords from A^n will have d_{min}/n satisfying the asymptotic Gilbert bound. This result means that virtually all binary codes are as good as the Gilbert bound, however, no one has so far proved the existence of binary codes better than promised by the Gilbert bound for very large n. In 1982, for all $q \geq 49$ and rates R in an interval depending on q, it was shown (Tsfasman, Vladut, and Zing 1982) that there exist arbitrarily long q-ary linear codes, in the class of Goppa codes, whose d_{min}/n exceeds the corresponding asymptotic Gilbert bound.

A.3   Gilbert bound for linear codes

Gilbert's construction in general will not yield a linear code, even when the alphabet A is GF(q). However, the idea behind the original construction can be adapted to cover the linear case. Given a field GF(q) and integers n and d, $2 \leq d \leq n$, the generator matrix $\mathbf{G}$ of a linear code with $d_{min} \geq d$ can be constructed as follows. Beginning with a list of all distinct q^n n-tuples, one should delete all those n-tuples in the Hamming sphere of radius $d - 1$ centered at $\mathbf{0}$. This means discarding all n-tuples of weight less than d. Choose for the first row, $\mathbf{g}_1$, of $\mathbf{G}$ any of the remaining n-tuples, then delete all n-tuples in the $q-1$ Hamming spheres of radius $d-1$ centered at the n-tuples $a_1\mathbf{g}_1$, $a_1 \neq 0$, $a_1 \in$ GF(q). Choose the second row of $\mathbf{G}$, $\mathbf{g}_2$, as any of the remaining n-tuples and delete all n-tuples in the $q(q-1)$ Hamming spheres of radius $d - 1$ centered at the n-tuples $a_1\mathbf{g}_1 + a_2\mathbf{g}_2$, $a_2 \neq 0$, $a_2 \in$ GF(q). This process of selection of n-tuples is continued until all n-tuples have been deleted. Since row $\mathbf{g}_i$ is chosen after q^{i-1} (possibly overlapping) Hamming spheres have been deleted, it follows that the resulting linear code satisfies the following bound, known as the Gilbert bound for linear codes.

Gilbert bound for linear codes: For any positive integers n and d, $2 \leq d \leq n$, there exists a q-ary (n, k) linear code with $d_{min} \geq d$, whose dimension k satisfies

$$S_{d-1} \geq q^{n-k}. \tag{A.2}$$

It is observed that the Gilbert bound for linear codes generally guarantees better codes than its counterpart in (A.1). This results because, for linear codes, the rate R has the form k/n. Expressions (A.1) and (A.2) are, however, formally identical because $k = Rn$.

Appendix B
MacWilliams' Identity for Linear Codes

B.1 Introduction

The MacWilliams' identity for linear codes relates the Hamming weight distribution of a code to the Hamming weight distribution of its associated dual code. The following derivation is based on probabilistic arguments and was first presented in (Chang and Wolf 1980). It makes use of the probability of undetected error, which is calculated in two different ways, and then the two results are equated. Applying the same transformation to the Hamming distance distribution of a nonlinear code, one obtains a set of nonnegative numbers with interesting interpretations in some cases.

B.2 The binary symmetric channel

The binary symmetric channel (BSC) with crossover probability ε, $0 \leq \varepsilon \leq 1$, is the binary-input binary-output channel where each input binary digit independently has probability $1 - \varepsilon$ of being correctly received and probability ε of being received in error. In a BSC with crossover probability ε, the probability of a particular pattern $\mathbf{x}$ of t errors in a block of n digits is given by

$$P(\mathbf{x}) = \varepsilon^t (1 - \varepsilon)^{n-t}, \ 0 \leq \varepsilon \leq 1.$$

B.3 Binary linear codes and error detection

Let $\mathbf{x}$ be a binary n-component vector with Hamming weight $w(\mathbf{x}) = t$. Consider an (n, k, d) binary linear code V with parity-check matrix $\mathbf{H}$. The syndrome associated to $\mathbf{x}$, denoted by $\mathbf{S} = \mathbf{S}(\mathbf{x})$, is defined by $\mathbf{S} = \mathbf{H}\mathbf{x}^{\mathrm{T}}$, where $\mathbf{x}^{\mathrm{T}}$ denotes the column vector which is the transpose of $\mathbf{x}$. The syndrome is zero, i.e., $\mathbf{S}(\mathbf{x}) = 0$, if and only if $\mathbf{x}$ is a codeword. An expanded parity-check matrix $\mathbf{H}^*$ is now defined which has as rows all the vectors in the row space of $\mathbf{H}$. Therefore $\mathbf{H}^*$ has 2^{n-k} rows. The rows of $\mathbf{H}^*$ are the codewords of the dual code of V. Associated to any binary n-vector $\mathbf{x}$ the expanded syndrome $\mathbf{S}^* = \mathbf{S}^*(\mathbf{x})$ is defined as $\mathbf{S}^* = \mathbf{H}^*\mathbf{x}^{\mathrm{T}}$.

LEMMA B.1 $\mathbf{S}^*(\mathbf{x}) = 0$ if and only if $\mathbf{S}(\mathbf{x}) = 0$. If $\mathbf{S}^*(\mathbf{x}) \neq 0$ then it contains half zeros and half ones among its 2^{n-k} components.

Proof: The proof follows by noticing that the components of $\mathbf{S}^*$ are obtained by taking all linear combinations of the components in $\mathbf{S}$. $\square$

If a binary (n, k, d) code V is used only to provide error detection on a BSC, an undetected error will occur if and only if the received word $\mathbf{r}$ is a codeword $\mathbf{v}'$ different from the transmitted codeword $\mathbf{v}$.

Letting A_i, $0 \le i \le n$, denote the number of codewords in V of Hamming weight i, i.e., denote the code weight distribution, the probability of an undetected error, P_{ue}, is given by

$$P_{\mathrm{ue}} = \sum_{t=1}^{n} A_t \varepsilon^t (1 - \varepsilon)^{n-t} = (1 - \varepsilon)^n \sum_{t=1}^{n} A_t [\varepsilon/(1 - \varepsilon)]^t.$$

The weight enumerator of the code is defined as the polynomial

$$A(z) = \sum_{t=0}^{n} A_t z^t,$$

and, since $A_0 = 1$ one can write P_{ue} as

$$P_{\mathrm{ue}} = (1 - \varepsilon)^n A[\varepsilon/(1 - \varepsilon)] - (1 - \varepsilon)^n. \tag{B.1}$$

This is one way of calculating P_{ue}. The dual code weight distribution will now be used to calculate P_{ue}.

Let the dual code be denoted by $V^\perp = U = \{\mathbf{u}_1, \mathbf{u}_2, \dots, \mathbf{u}_{2^{n-k}}\}$. Let $\mathbf{r}$ denote a received binary n-tuple and let F_i denote the event $\mathbf{r}\mathbf{u}_i^{\mathrm{T}} = 1$, i.e., the event that $\mathbf{u}_i$ detects an odd number of errors in $\mathbf{r}$. If $w(\mathbf{u}_i) = w_i$, then $P(F_i)$ is the probability of the occurrence of an odd number of errors, caused by the BSC, in the w_i nonzero positions of $\mathbf{u}_i$. Thus,

$$P(F_i) = \sum_{j=1}^{w_i} \binom{w_i}{j} \varepsilon^j (1 - \varepsilon)^{w_i - j}, \text{ for } j \text{ odd.}$$

The expression for $P(F_i)$ can be written as (see Section B.6)

$$P(F_i) = (1/2)[1 - (1 - 2\varepsilon)^{w_i}].$$

As a consequence of *Lemma B.1*, either none of the events F_i occur or exactly 2^{n-k-1} such events occur. Therefore,

$$\sum_{i=1}^{2^{n-k}} P(F_i) = 2^{n-k-1} P(\cup_{i=1}^{2^{n-k}} F_i),$$

where

$$P(\cup_{i=1}^{2^{n-k}} F_i) = P(\mathbf{S}^* \ne 0),$$

i.e., the summation on the left-hand side is equal to 2^{n-k-1} times the probability of a nonzero syndrome. Thus, one can write

$$P_{\mathrm{ue}} = 1 - (1 - \varepsilon)^n - P(\mathbf{S}^* \ne 0) =$$

$$1 - (1 - \varepsilon)^n - (1/2^{n-k-1}) \sum_{i=1}^{2^{n-k}} (1/2)[1 - (1 - 2\varepsilon)^{w_i}],$$

since $(1 - \varepsilon)^n$ is the probability of a successful transmission. Let B_t, $0 \le t \le n$, denote the number of codewords of Hamming weight t in the dual code $V^{\perp}$. The weight enumerator of $V^{\perp}$ is written as

$$B(z) = \sum_{t=0}^{n} B_t z^t.$$

Thus,

$$P_{\text{ue}} = 1 - (1 - \varepsilon)^n - (1/2^{n-k}) \sum_{i=1}^{2^{n-k}} [1 - (1 - 2\varepsilon)^{w_i}]$$

or,

$$P_{\text{ue}} = 2^{-n+k} \sum_{i=1}^{2^{n-k}} (1 - 2\varepsilon)^{w_i} - (1 - \varepsilon)^n.$$

However,

$$B(1 - 2\varepsilon) = \sum_{t=0}^{n} B_t (1 - 2\varepsilon)^t = \sum_{i=1}^{2^{n-k}} (1 - 2\varepsilon)^{w_i},$$

and thus

$$P_{\text{ue}} = 2^{-n+k} B(1 - 2\varepsilon) - (1 - \varepsilon)^n. \tag{B.2}$$

Equating (B.1) and (B.2) and making $z = 1 - 2\varepsilon$ the desired formula results

$$B(z) = 2^{-k}(1 + z)^n A[(1 - z)/(1 + z)], \tag{B.3}$$

which is known as MacWilliams' identity for binary linear codes.

The weight enumerator $A(z)$ of the (n, k, d) linear binary code V uniquely determines $B(z)$, the weight enumerator of the dual code $V^{\perp}$.

EXAMPLE B.2 *Consider the binary $(7, 4, 3)$ Hamming code with the following parity-check matrix:*

$$\mathbf{H} = \begin{bmatrix} 0 & 0 & 0 & 1 & 1 & 1 & 1 \\ 0 & 1 & 1 & 0 & 0 & 1 & 1 \\ 1 & 0 & 1 & 0 & 1 & 0 & 1 \end{bmatrix} = \begin{bmatrix} \mathbf{u}_1 \\ \mathbf{u}_2 \\ \mathbf{u}_3 \end{bmatrix}.$$

The extended parity-check matrix $\mathbf{H}^*$ is thus

$$\mathbf{H}^* = \begin{bmatrix} 0 & 0 & 0 & 0 & 0 & 0 & 0 \\ 0 & 0 & 0 & 1 & 1 & 1 & 1 \\ 0 & 1 & 1 & 0 & 0 & 1 & 1 \\ 1 & 0 & 1 & 0 & 1 & 0 & 1 \\ 0 & 1 & 1 & 1 & 1 & 0 & 0 \\ 1 & 0 & 1 & 1 & 0 & 1 & 0 \\ 1 & 1 & 0 & 0 & 1 & 1 & 0 \\ 1 & 1 & 0 & 1 & 0 & 0 & 1 \end{bmatrix}.$$

Let $\mathbf{r}$ be the received n-tuple. Then one has $\mathbf{S} = \mathbf{H}\mathbf{r}^{\text{T}}$, or

$$\mathbf{S} = [\mathbf{u}_1\mathbf{r}^{\text{T}}, \mathbf{u}_2\mathbf{r}^{\text{T}}, \mathbf{u}_3\mathbf{r}^{\text{T}}].$$

It follows that for $\mathbf{S}^*$ one obtains

$$\mathbf{S}^* = [\mathbf{u}_1\mathbf{r}^{\text{T}}, \mathbf{u}_2\mathbf{r}^{\text{T}}, \dots, \mathbf{u}_{2^{n-k}}\mathbf{r}^{\text{T}}],$$

where $\mathbf{S}^* \ne 0$ if and only if $\mathbf{S} \ne 0$.

Now let $\mathbf{r} = (0, 1, 0, 1, 0, 0, 0)$, *then one obtains* $\mathbf{S} = [1, 1, 0]^{\mathrm{T}}$ *and* $\mathbf{S}^* = [0, 1, 1, 0, 0,$
$1, 1, 0]^{\mathrm{T}}$ *with half ones and half zeros. Finally, since the weight distribution of the*
(7,4,3) Hamming code is

$$A(z) = 1 = 7z^3 + 7z^4 + z^7,$$

for its dual code, i.e., the m-sequence (7,3,4) code, one obtains through the application
of MacWilliams' identity the following expression:

$$B(z) = 2^{-4}(1 + z)^7 A[(1 - z)/(1 + z)],$$

which after being expanded and simplified leads to

$$B(z) = 1 + 7z^4.$$

B.4   The q-ary symmetric channel

The q-ary symmetric channel with crossover probability $\varepsilon/(q - 1)$, $0 \le \varepsilon \ne 1$,
is the q-ary-input q-ary-output channel where each input symbol independently has
probability $1 - \varepsilon$ of being correctly received and probability $\varepsilon/(q-1)$ of being received
in error. The probability of a particular pattern $\mathbf{x}$ of t errors in a block of length n
digits is given by

$$P(\mathbf{x}) = [\varepsilon/(q - 1)]^t (1 - \varepsilon)^{n-t}, \ 0 \le \varepsilon \le 1.$$

B.5   Linear codes over GF(q)

For linear (n, k, d) codes over GF(q), the MacWilliams' identity has the form

$$B(z) = q^{-k}[1 + (q - 1)z]^n A[(1 - z)/(1 + (q - 1)z)]. \tag{B.4}$$

For the derivation of (B.4), using the probability of undetected error argument, we
use the respective definitions given earlier for both the extended parity-check matrix
and the extended syndrome. For any received vector $\mathbf{x}$, with coordinates in GF(q),
it is easy to show that the extended syndrome is either all-zero or contains each
element of GF(q) occurring the same number $(q-1)q^{n-k-1}$ of times. The probability
of an undetected error, denoted by P_{ue}, can be written in terms of the code weight
distribution A_i and weight enumerator $A(z)$ as

$$P_{\mathrm{ue}} = \sum_{t=1}^{n} A_t (1 - \varepsilon)^{n-t} [\varepsilon/(q - 1)]^t,$$

i.e.,

$$P_{\mathrm{ue}} = (1 - \varepsilon)^n A[\varepsilon/(q - 1)(1 - \varepsilon)] - (1 - \varepsilon)^n, \tag{B.5}$$

since $A_0 = 1$.

Let F_i denote the event $\mathbf{r}\mathbf{u}_i^{\mathrm{T}} \ne 0$, i.e., the event that $\mathbf{u}_i$ detects one or more errors
in $\mathbf{r}$. By a reasoning similar to the one used in the binary case, we have

$$\sum_{i=1}^{q^{n-k}} P(F_i) = (q - 1)q^{n-k-1} P(\cup_{i=1}^{q^{n-k}} F_i),$$

where

$$P(\cup_{i=1}^{q^{n-k}} F_i) = P(\mathbf{S}^* \ne 0).$$

The probability of an undetected error can be written as

$$P_{ue} = 1 - (1-\varepsilon)^n - P(\mathbf{S}^* \neq 0) =$$

$$P_{ue} = 1 - (1-\varepsilon)^n - \left[\frac{1}{(q-1)q^{n-k-1}}\right] \sum_{i=1}^{q^{n-k}} P(F_i).$$

If $w(\mathbf{u}_i) = w_i$, then $P(F_i)$ is the probability of the occurrence of a parity-check failure due to errors, caused by the q-ary symmetric channel, in the w_i nonzero positions of $\mathbf{u}_i$.

LEMMA B.3 *The expression for $P(F_i)$ is given by*

$$P(F_i) = \frac{(q-1)}{q}\left[1 - \left(1 - \frac{\varepsilon q}{q-1}\right)^{w_i}\right].$$

We give a proof of this lemma in Section B.6.
Thus one way of computing P_{ue} is through the formula

$$P_{ue} = 1 - (1-\varepsilon)^n - P(\mathbf{S}^* \neq 0) =$$

$$1 - (1-\varepsilon)^n - \frac{1}{q^{n-k-1}} \sum_{i=1}^{q^{n-k}} \frac{1}{q}\left[1 - \left(1 - \frac{\varepsilon q}{q-1}\right)^{w_i}\right]$$

or

$$P_{ue} = q^{-n+k} \sum_{i=1}^{q^{n-k}} \left(1 - \frac{\varepsilon q}{q-1}\right)^{w_i} - (1-\varepsilon)^n.$$

Let $B(z)$ and B_t, $0 \leq t \leq n$, denote respectively the weight enumerator and the weight distribution of the dual code. Thus,

$$B(z) = \sum_{t=0}^{n} B_t z^t.$$

However,

$$B\left(1 - \frac{\varepsilon q}{q-1}\right) = \sum_{t=0}^{n} B_t \left(1 - \frac{\varepsilon q}{q-1}\right)^t = \sum_{i=1}^{q^{n-k}} \left(1 - \frac{\varepsilon q}{q-1}\right)^{w_i},$$

and we can write P_{ue} in terms of $B(.)$ as

$$P_{ue} = q^{-n+k} B\left(1 - \frac{\varepsilon q}{q-1}\right) - (1-\varepsilon)^n. \qquad (B.6)$$

Equating (B.5) and (B.6) we obtain

$$(1-\varepsilon)^n A\left[\frac{\varepsilon}{(q-1)(1-\varepsilon)}\right] = q^{-n+k} B\left(1 - \frac{\varepsilon q}{q-1}\right), \qquad (B.7)$$

and finally, letting $z = 1 - \frac{\varepsilon q}{q-1}$ in (B.7) we obtain

$$B(z) = q^{-k}[1 + z(q-1)^n]A\left(\frac{1-z}{1+z(q-1)}\right), \qquad (B.8)$$

which represents the MacWilliams' identity for q-ary linear codes.

B.6   The binomial expansion

Consider the expansion of the following binomials:

$$(a+b)^n = \sum_{i=0}^{n} \binom{n}{i} b^i a^{n-i} \tag{B.9}$$

$$(a-b)^n = \sum_{i=0}^{n} (-1)^i b^i a^{n-i}. \tag{B.10}$$

An expression for the sum of the terms where i is an odd number is obtained by subtracting (B.10) from (B.9), i.e.,

$$(a+b)^n - (a-b)^n = 2\sum_{i=1}^{n} \binom{n}{i} b^i a^{n-i}, \text{ for } i \text{ odd.} \tag{B.11}$$

Substituting $a = 1 - \varepsilon$ and $b = \varepsilon$ in (B.11), it follows that

$$1 - (1 - 2\varepsilon)^{w_i} = 2P(F_i),$$

which is the desired expression for $P(F_i)$ in the binary case.

Proof: (**of Lemma B.3**) To visualize how a parity-check failure occurs, in the q-ary case, let us draw a tree diagram as follows. Start with a single node which we call node zero. From node zero we draw $q - 1$ branches and explain how the tree grows by concentrating our description on one of these branches. Take any nonzero element of GF(q), call it a_1. Then label this chosen branch with $B_{11} = a_1 u_{i1}$, which is the product of a_1 with the first nonzero component of $\mathbf{u}_i$. Label the node at the end of B_{11} with $\mathbf{S}_{11} = B_{11}$. In the sequel, we denote the nonzero elements of $\mathbf{u}_i$ by u_{ij}, $1 \leq j \leq w_i$. Now extend the tree, from node $\mathbf{S}_{11}$ by drawing $q - 1$ branches, corresponding to the nonzero elements of GF(q) as follows. Branch j, $1 \leq j \leq q-1$, stemming from $\mathbf{S}_{11}$, is labeled with the sum $B_{2j} = a_1 u_{i1} + a_j u_{i2}, a_j \neq 0, a_j \in$ GF(q). Obviously, only one of these sums will be zero because the congruence $a_1 u_{i1} + a_j u_{i2} \equiv 0$ mod q, has a unique solution in the unknown variable $a_j \neq 0$ $a_j \in$ GF(q). Each branch B_{2j} leads to a node $\mathbf{S}_{2j}$, labeled as $\mathbf{S}_{2j} = B_{2j}$, $1 \leq j \leq q - 1$. The tree is again extended by drawing $q - 1$ branches from each node $\mathbf{S}_{2j}$, $1 \leq j \leq q - 1$. Each new branch is now labeled with the sum B_{3j}, $1 \leq j \leq q - 1$, defined in a way analogous to the previous sums. Now, however, we observe the important fact that the node for which $\mathbf{S}_{2j} = 0$ can only lead to branches with $B_{3j} \neq 0$. This process of expanding the tree is continued until we have used all the nonzero components of $\mathbf{u}_i$. Since a_1 can have $q - 1$ nonzero values, the following tree evolution rule is observed, considering what occurs at the end of the kth step.

(a) The total number of nodes is $(q - 1)^k$, $k \geq 1$.

(b) Each of the nodes satisfying the parity-check equation leads to $q - 1$ branches, none of which satisfy that parity-check equation.

(c) Each of the nodes nonsatisfying the parity-check equation leads to $q-1$ branches, only one of which satisfies that parity-check equation.

Denote by N_k the number of nodes that do not satisfy the parity-check equation, at the end of the kth step. It follows from the tree evolution rule that

$$N_k = (q-1)^k - N_{k-1}, \ k \geq 1,$$

which can be solved, by defining $N_0 = 1$, to give

$$N_k = \sum_{e=0}^{k-1}(q-1)^{k-e}(-1)^e.$$

Now one can write $P(F_i)$ as

$$
\begin{aligned}
P(F_i) &= \sum_{s=1}^{w_i}\binom{w_i}{s}(1-\varepsilon)^{w_i-s}[\varepsilon/(q-1)]^s N_s \\
&= \sum_{s=1}^{w_i}\binom{w_i}{s}\varepsilon^s(1-\varepsilon)^{w_i-s}\sum_{l=0}^{s-1}[-1/(q-1)]^l,
\end{aligned}
$$

however,

$$\sum_{l=0}^{s-1}[-1/(q-1)]^l = \frac{1-[-1/(q-1)]^s}{q/(q-1)}$$

thus,

$$
\begin{aligned}
P(F_i) &= \frac{(q-1)}{q}\sum_{s=1}^{w_i}\binom{w_i}{s}\varepsilon^s(1-\varepsilon)^{w_i-s}\{1-[-1/(q-1)]^s\} \\
&= \frac{(q-1)}{q}\left[\sum_{s=1}^{w_i}\binom{w_i}{s}\varepsilon^s(1-\varepsilon)^{w_i-s} - \sum_{s=1}^{w_i}\binom{w_i}{s}\left(\frac{-\varepsilon}{q-1}\right)^s(1-\varepsilon)^{w_i-s}\right] \\
&= \frac{(q-1)}{q}\left[\sum_{s=0}^{w_i}\binom{w_i}{s}\varepsilon^s(1-\varepsilon)^{w_i-s} - \sum_{s=0}^{w_i}\binom{w_i}{s}\left(\frac{-\varepsilon}{q-1}\right)^s(1-\varepsilon)^{w_i-s}\right] \\
&= \frac{(q-1)}{q}\left[1-\left(1-\frac{\varepsilon q}{q-1}\right)^{w_i}\right].
\end{aligned}
$$

$\square$

B.7 Digital transmission using N regenerative repeaters

In this section we observe that the expression for $P(F_i)$ is precisely the same as that for the probability of error in digital transmission, using N regenerative repeaters, in a q-ary symmetric channel. Consider the transmission of a voltage level a, where a can assume any one of q distinct values, through a cascade of N regenerative repeaters. We assume that the noise between consecutive repeaters is of the same type, no matter which pair of consecutive repeaters we choose, and that the associated channel can be modeled as a q-ary symmetric channel. Due to the channel symmetry, the value of $P(F_i)$ is not dependent on the actual value of the level a.

Appendix C
Frequency Domain Decoding Tools

C.1   Finite Field Fourier Transform

DEFINITION C.1 *Let* $\mathbf{v} = (v_0, v_1, \ldots, v_i, \ldots, v_{n-1})$ *be an n-tuple with coefficients in* $\mathrm{GF}(q)$, *where n divides* $q^m - 1$ *for some positive integer m, and let* α *be an element of multiplicative order n in* $\mathrm{GF}(q^m)$. *The n-tuple* $\mathbf{V} = (V_0, V_1, \ldots, V_j, \ldots, V_{n-1})$ *defined over* $\mathrm{GF}(q^m)$, *whose components are given by:*

$$V_j = \sum_{i=0}^{n-1} \alpha^{ij} v_i, \quad j = 0, 1, \ldots, n-1,$$

is called the finite field Fourier transform of $\mathbf{v}$.

Vectors $\mathbf{v}$ and $\mathbf{V}$ constitute a Fourier transform pair, usually represented as:

$$\mathbf{v} \Longleftrightarrow \mathbf{V}$$

or

$$\{v_i\} \longleftrightarrow \{V_j\}.$$

In conformity with the terminology employed in electrical engineering for the conventional Fourier transform, indices i and j are referred to as time and frequency, respectively. As a consequence of Definition C.1 some important properties of the finite field Fourier transform are derived which have great utility in algebraic coding, and are presented next.

PROPERTY 1. Over $\mathrm{GF}(q)$, a finite field of characteristic p, the components v_i, $0 \le i \le n-1$, of a vector $\mathbf{v}$ are related to the components V_j, $0 \le j \le n-1$, of its finite field Fourier transform $\mathbf{V}$ through the expressions:

$$V_j = \sum_{i=0}^{n-1} \alpha^{ij} v_i$$

$$v_i = \frac{1}{(n \mod p)} \sum_{j=0}^{n-1} \alpha^{-ij} V_j.$$

PROPERTY 2. Given a Fourier transform pair $\{v_i\} \leftrightarrow \{V_j\}$ and constants c, i_0 and k, then:

(a)	$c\{v_i\}$	$\longleftrightarrow$	$c\{V_j\}$	(Linearity)
(b)	$\{v_{i-i_0}\}$	$\longleftrightarrow$	$\{V_j\alpha^{ji_0}\}$	(Time shift)
(c)	$\{v_{ki}\}$	$\longleftrightarrow$	$\{V_{j/k}\}$	(Scaling)$(k,n) = 1$.

PROPERTY 3. Given two Fourier transform pairs

$$\{f_i\} \longleftrightarrow \{F_j\}$$

and

$$\{g_i\} \longleftrightarrow \{G_j\},$$

then:

$$\{f_i g_i\} \longleftrightarrow \{(1/n)F_k G_{j-k}\} \text{ (Frequency domain convolution).}$$

Vectors **v** and **V** are also usually represented by polynomials as follows:

$$\begin{aligned} v(x) &= v_{n-1}x^{n-1} + v_{n-2}x^{n-2} + \cdots + v_1 x + v_0 \\ V(z) &= V_{n-1}z^{n-1} + V_{n-2}z^{n-2} + \cdots + V_1 z + V_0. \end{aligned}$$

It follows from this polynomial representation of **v** and **V** that

$$V_j = \sum_{i=0}^{n-1} \alpha^{ij} v_i = v(\alpha^j)$$

$$v_i = (1/n)\sum_{j=0}^{n-1} \alpha^{-ij} V_j = (1/n)V(\alpha^{-i}).$$

In this manner, we obtain a relationship between the roots of a polynomial in one domain and the components of the corresponding finite field Fourier transform pair in the other domain. In other words, $v(x)$ has a root α^j, i.e., $v(\alpha^j) = 0$, if and only if $V_j = 0$. Conversely, $V(z)$ has a root α^{-i}, i.e., $V(\alpha^{-i}) = 0$, if and only if $v_i = 0$.

C.2   The Euclidean algorithm

The Euclidean algorithm (Clark and Cain 1981, p.195) is a technique which allows the calculation of the greatest common divisor of two integers, or of two polynomials. Our main interest is to solve the key equation for decoding cyclic codes, and not to calculate the greatest common divisor. Beginning with two polynomials, $a(z)$ and $b(z)$, the Euclidean algorithm employs the following relationship:

$$f_i(z)a(z) + g_i(z)b(z) = r_i(z),$$

where, for any one of the polynomials $f_i(z)$, $g_i(z)$, or $r_i(z)$ replacing $h_i(z)$, the following recurrence relation is applied:

$$h_i(z) = h_{i-2}(z) - q_i(z)h_{i-1}(z),$$

subject to the following initial conditions:

$$\begin{aligned} f_{-1}(z) &= g_0(z) &= 1 \\ f_0(z) &= g_{-1}(z) &= 0 \\ r_{-1}(z) &= a(z) \\ r_0(z) &= b(z). \end{aligned}$$

The polynomial $q_i(z)$ is given by the integer part with non-negative exponents of the quotient

$$r_{i-2}(z)/r_{i-1}(z).$$

To solve the key equation we consider

$$a(z) = z^{2t}$$

and

$$b(z) = S(z),$$

and apply the Euclidean algorithm, stopping when the degree of $r_i(z)$ is less than t. We then take

$$L(z) = g_i(z).$$

The Berlekamp–Massey algorithm (Berlekamp 1968, Massey 1969), described in Chapter 4, provides an alternative way for solving the key equation, slightly more efficient than the Euclidean algorithm. Massey (1969) treats the Berlekamp–Massey algorithm in a very thorough manner and allows the reader to understand the algorithm in terms of a generalized sequence synthesis procedure.

Bibliography

A, Nguyen Q., L. Györfi, and J. L. Massey. 1992. "Constructions of binary constant-weight cyclic codes and cyclically permutable codes." *IEEE Trans. on Information Theory* 38 (3): 940–949.

Adams, C. M., and H. Meijer. 1989. "Security-related comments regarding McEliece's public-key cryptosystem." *IEEE Trans. on Information Theory* 35 (2): 454–455.

Alencar, M. S. 2009. *Digital Television Systems*. New York: Cambridge University Press.

Berlekamp, E. 1973. "Goppa codes." *IEEE Trans. on Information Theory* 19 (5): 590–592.

Berlekamp, E. R. 1968. *Algebraic Coding Theory*. New York: McGraw-Hill.

Berlekamp, E., and O. Moreno. 1973. "Extended double-error-correcting binary Goppa codes are cyclic." *IEEE Trans. on Information Theory* 19 (6): 817–818.

Berrou, C., A. Glavieux, and P. Thitimajshima. 1993. "Near Shannon limit error-correcting coding and decoding: Turbo Codes." In *Proceedings of the 1993 IEEE International Conference on Communications (ICC'93)*, 367–377. New York: IEEE Press.

Blahut, R. E. 1983. *Theory and Practice of Error Control Codes*. Reading, Mass.: Addison Wesley.

Castagnoli, G., J. L. Massey, and P. A. Schoeller. 1991. "On repeated-root cyclic codes." *IEEE Trans. on Information Theory* 37 (2): 337–342.

Chang, S. C., and J. K. Wolf. 1980. "A simple derivation of the MacWilliams' identity for linear codes." *IEEE Trans. on Information Theory* 26 (4): 476–477.

Chen, C. L., and M. Y. Hsiao. 1984. "Error-correcting codes for semiconductor memory applications: A state of the art review." *IBM Jour. Res. and Dev.* 28:124–134.

Clark, G. C., and J. B. Cain. 1981. *Error-Correction Coding for Digital Communications.* New York: Plenum Press.

Cover, T. M., and J. A. Thomas. 2006. *Elements of Information Theory.* New Jersey: Second Edition, Wiley Interscience.

Daemen, J., and V. Rijmen. 2002. *The Design of RijndaeL: AES - The Advanced Encryption Standard.* New York: Springer.

Denning, D. E. R. 1982. *Cryptography and Data Security.* Addison Wesley.

Gallager, R. G. 1963. *Low-Density Parity Check Codes.* Cambridge, Mass.: MIT Press.

Goppa, V. D. 1970. "A new class of error-correcting codes." *Problems of Information Transmission* 6 (3): 207–212.

Hamming, R. W. 1950. "Error detecting and error correcting codes." *Bell Syst. Tech. Journal* 49:147–160.

Hammons Jr., A. R., P. V. Kumar, A. R. Calderbank, N. J. A. Sloane, and P. Solé. 1994. "The Z4-linearity of Kerdock, Preparata, Goethals, and related codes." *IEEE Trans. on Information Theory* 40 (2): 301–319.

Hartmann, C. R. P., and L. D. Rudolph. 1976. "An optimum symbol-by-symbol decoding rule for linear codes." *IEEE Trans. on Information Theory* 22 (5): 514–517.

Honary, B., and G. Markarian. 1998. *Trellis Decoding of Block Codes.* London: Kluwer Academic Publishers.

Immink, K. A. S. 1994. *RS codes and the compact disk.* Edited by Steve Wicker and Vijay Bhargava. New York: IEEE Press.

Konheim, A. G. 1981. *Cryptography a Primer.* John Wiley & Sons.

Kou, Y., S. Lin, and M. Fossorier. 2001. "Low density parity check codes based on finite geometries: a rediscovery and new results." *IEEE Trans. on Information Theory* 47 (7): 2711–2736.

Krouk, E. 1993. "A new public-key cryptosystem." In *Proceedings of the 6th Joint Swedish-Russian International Workshop on Information Theory,* 285–286. Sweden: Lund Studentlitteratur.

Lidl, R., and H. Niederreiter. 2006. *Introduction to Finite Fields and their Applications.* Cambridge: Cambridge University Press.

Lin, Shu, and Daniel J. Costello Jr. 2004. *Error Control Coding : Fundamentals and Applications.* New Jersey: Pearson Prentice-Hall.

Lint, J. H. van. 1982. *Introduction to Coding Theory.* New York: Springer Verlag.

MacKay, D. J. C. 1999. "Good error-correcting codes based on very sparse matrices." *IEEE Trans. on Information Theory* 45 (2): 399–432.

MacKay, D. J. C., and R. M. Neal. 1996. "Near Shannon limit performance of low density parity check codes." *Electronics Letters* 32 (18): 1645–1646.

MacWilliams, F. J., and N. J. A. Sloane. 1977. *The Theory of Error-Correcting Codes.* Amsterdam: North-Holland.

Massey, J. L. 1963. *Threshold Decoding.* Cambridge, Mass.: MIT Press.

———. 1969. "Shift-register synthesis and BCH decoding." *IEEE Trans. on Information Theory* 15 (5): 122–127.

———. 1985. *Handbook of Applicable Mathematics,* Chapter 16, vol. V, Part B, Combinatorics and Geometry, pp.623–676. Chichester and New York: Wiley.

———. 1998. *Cryptography: Fundamentals and Applications.* ETH-Zurich: Class Notes.

McEliece, R. J. 1978. "A public-key cryptosystem based on algebraic coding theory." *DSN Progress Report, Jet Propulsion Laboratory* 42 (44): 42–44.

———. 1987. *Finite Fields for Computer Scientists and Engineers.* Lancaster: Kluwer Academic Publishers.

Moreira, J. C., and P. G. Farrell. 2006. *Essentials of Error-Control Codes.* West Sussex: John Wiley & Sons, Ltd.

Parkinson, B. W., and J. J. Spilker Jr. 1996. *Global Positioning System: Theory and Applications.* New York: American Institute of Aeronautics/Astronautics.

Patterson, N. J. 1975. "The algebraic decoding of Goppa codes." *IEEE Trans. on Information Theory* 21 (2): 203–207.

Peterson, W. W., and Edward J. Weldon Jr. 1972. *Error-Correcting Codes.* MIT Press.

Pless, V. 1982. *An Introduction to the Theory of Error-Correcting Codes.* New York: John Wiley & Sons., Inc.

Rao, T. R. N., and K. -H. Nam. 1989. "Private-key algebraic-code encryptions." *IEEE Trans. on Information Theory* 35 (4): 829–833.

Rocha Jr., V. C. da. 1993. *Some protocol sequences for the M-out-of-T-sender collision channel without feedback.* Edited by Bahram Honary, Michael Darnell, and Patrick Farrell. Lancaster: H & W Communications Limited.

Rocha Jr., V. C. da, and D. L. Macedo. 1996. "Cryptanalysis of Krouk's public-key cypher." *Electronics Letters* 32 (14): 1279–1280.

Shannon, C. E. 1948. "A mathematical theory of communication." *Bell System Technical Journal* 27:379–423, 623–656.

———. 1949. "Communication theory of secrecy systems." *Bell System Technical Journal* 27:656–715.

Tanner, M. 1981. "A recursive approach to low complexity codes." *IEEE Trans. on Information Theory* 27 (5): 533–547.

Tsfasman, M. A., S. G. Vladut, and Th. Zing. 1982. "Modular curves, Shimura curves, and Goppa codes better than the Varshamov-Gilbert bound." *Math. Nachr.* 104:13–28.

Vasil'ev, Yu. L. 1962. "Nongroup close-packed codes." *Prob. Cybernet.* 8:337–339.

Viterbi, A. J., and J. K. Omura. 1979. *Principles of Digital Communication and Coding.* New York: McGraw-Hill Book Company.

Wolf, J. K. 1978. "Efficient maximum likelihood decoding of linear block codes using a trellis." *IEEE Trans. on Information Theory* 24 (1): 76–80.

About the Author

Valdemar Cardoso da Rocha Jr. was born in Jaboatão, Pernambuco, Brazil, on August 27, 1947. He received the B.Sc. degree in Electrical/Electronics Engineering from the Escola Politécnica, Recife, Brazil, in 1970, and the Ph.D. degree in Electronics from the University of Kent at Canterbury, England, in 1976. In 1976, he joined the faculty of the Federal University of Pernambuco, Recife, Brazil, as an Associate Professor and founded the Electrical Engineering Postgraduate Programme. From 1992 to 1996 he was Head of the Department of Electronics and Systems and in 1993 became Professor of Telecommunications.

He has often been a consultant to both the Brazilian Ministry of Education and the Ministry of Science and Technology on postgraduate education and research in electrical engineering. For two terms (1993–1995 and 1999–2001) he was the Chairman of the Electrical Engineering Committee in the Brazilian National Council for Scientific and Technological Development. From 1990 to 1992, he was a visiting professor at the Swiss Federal Institute of Technology-Zurich, Institute for Signal and Information Processing. In 2005 and 2006, he was a visiting professor at the University of Leeds; and during 2007, he was a visiting professor at Lancaster University.

He is a founding member, former President (2004–2008), and Emeritus Member (2008) of the Brazilian Telecommunications Society. He is also a Life Senior Member (2013) of the IEEE Communications Society and the IEEE Information Theory Society and a Fellow (1992) of the Institute of Mathematics and its Applications. His research interests include information theory, error-correcting codes, and cryptography.

Index

FORTHCOMING TITLES IN OUR COMMUNICATIONS AND SIGNAL PROCESSING COLLECTION

Orlando Baiocchi, University of Washington Tacoma, Editor

Information Theory By Marcelo S. Alencar

Signal Integrity: The Art of Interconnect Design For High Speed and High Reliability Circuits By Joel Jorgenson, PhD

Cryptography Explained By Raj Katti

Not only is Momentum Press actively seeking collection editors for Collections, but the editors are also looking for authors. For more information about becoming an MP author, please go to http://www.momentumpress.net/contact!

Announcing Digital Content Crafted by Librarians

Momentum Press offers digital content as authoritative treatments of advanced engineering topics, by leaders in their fields. Hosted on ebrary, MP provides practitioners, researchers, faculty and students in engineering, science and industry with innovative electronic content in sensors and controls engineering, advanced energy engineering, manufacturing, and materials science. **Momentum Press offers library-friendly terms:**

- perpetual access for a one-time fee
- no subscriptions or access fees required
- unlimited concurrent usage permitted
- downloadable PDFs provided
- free MARC records included
- free trials

The **Momentum Press** digital library is very affordable, with no obligation to buy in future years.

For more information, please visit **www.momentumpress.net/library** or to set up a trial in the US, please contact **mpsales@globalepress.com.**